Biology, Physiology and Molecular Biology of Weeds

Biology, Physiology and Molecular Biology of Weeds

Editor

Mithila Jugulam
Department of Agronomy
Kansas State University
Manhattan, Kansas
USA

 **CRC Press**
Taylor & Francis Group
Boca Raton London New York

CRC Press is an imprint of the
Taylor & Francis Group, an **informa** business

A SCIENCE PUBLISHERS BOOK

CRC Press
Taylor & Francis Group
6000 Broken Sound Parkway NW, Suite 300
Boca Raton, FL 33487-2742

First issued in paperback 2020

ISBN-13: 978-1-4987-3711-1 (hbk)
ISBN-13: 978-0-367-78214-6 (pbk)

Library of Congress Cataloging-in-Publication Data

Names: Jugulam, Mithila, 1964-
Title: Biology, physiology, and molecular biology of weeds / [edited by] Mithila Jugulam, Department of Agronomy, Kansas State University, Manhattan, Kansas, USA.
Description: Boca Raton, FL : CRC Press, 2017. | "A science publishers book." | Includes bibliographical references and index.
Identifiers: LCCN 2017004899| ISBN 9781498737111 (hardback) | ISBN 9781498737128 (e-book)
Subjects: LCSH: Weeds. | Weeds--Control. | Herbicides.
Classification: LCC SB611 .B5587 2017 | DDC 632/.5--dc23
LC record available at https://lccn.loc.gov/2017004899

Visit the Taylor & Francis Web site at
http://www.taylorandfrancis.com

and the CRC Press Web site at
http://www.crcpress.com

Preface

Although relatively new, weed science is an important and integral discipline of agronomy. The innovations and advancements made in this area have had a major impact on crop production worldwide. Until the 19th century, growers were primarily dependent on cultural or mechanical weed control tactics. With the discovery of synthetic herbicides, chemical weed control has become essential and indispensable in agriculture. Along with the rapid and increased adoption of chemical weed control, tremendous progress was made towards a better understanding of biology and ecology of weeds, which contributed significantly for the effective weed management, thereby, exceptional progress in agricultural productivity. While agricultural losses due to weed competition have minimized, as a result of the adoption of chemical weed control, herbicide selection has resulted in an evolution of weed resistance to herbicides.

In the mid-1900s with the introduction of herbicide-tolerant technology, weed control has become even more efficient in many cropping systems. Quick and rapid adoption of this technology in countries like the USA, Australia, Argentina and Brazil, has resulted in an extensive and exclusive use of, a single mode of action of herbicides such as glyphosate. Consequently, a number of economically important weed species have also evolved a resistance to glyphosate, which then threatens the sustainable use of this herbicide for weed management in many countries. Currently, weed resistance to herbicides is one of the burning topics in the agriculture which has already gained significant attention of people ranging from farming communities, to academia and policy makers.

A number of books in weed science have been published in weed biology, ecology, and management. In this book an effort has been made to comprehensively discuss the new developments in weed biology, ecology, physiology and molecular biology aspects to help readers access the most recent and important advancements made in these fields.

This book is divided into 10 chapters. Chapter 1 introduces the information on the importance of weeds in agriculture and the history of weed control. Information about the early phases of research in weed science and how it has emerged as an important discipline of agriculture is also presented. Chapter 2 provides an overview of weed seed bank dynamics and its impact on weed infestation. An integrative approach that uses chemical, cultural, and mechanical practices to manage weed seed banks is also discussed. In Chapter 3, physiological and environmental factors affecting weed seed germination and emergence are reviewed. Critical information on seed biology and the dormancy of different weed species as well as their behavior in different environmental conditions is examined. Chapter 4 is a comprehensive review of chemical, cultural, and mechanical weed control strategies employed in different cropping systems around the world with emphasis on some of the latest integrative methods that can be used to achieve maximum control of problematic weed populations.

Weed resistance to herbicides is introduced in Chapter 5 and an overview of different phases of herbicide resistance evolution over the years is presented. In addition, mechanisms underlying herbicide resistance are discussed along with future research opportunities and new approaches to strategically manage herbicide resistance evolution. Chapter 6 describes the importance and potential of inter specific gene flow in the spread of herbicide resistance. In this context, herbicide tolerant traits of some of the major crop species, and the distribution and prevalence of their wild relatives are summarized. In Chapter 7, herbicide safeners and their role in crop selectivity are discussed in detail. The chapter focuses on the development of herbicide safeners for commercial use in chemical weed control technology. It also provides an in-depth and current knowledge on the physiological, biochemical, and molecular pathways involved in the mechanism of action of herbicide safeners. Chapter 8 provides the history and current status of metabolic resistance to herbicides in crops and weeds. The chapter reviews the role of cytochrome P450, glutathione-s-transferase, and other enzymes in metabolic resistance of crops as well as some of the recent advances in the study of not target-site resistance in different weed species.

Chapter 9 examines the mechanisms underlying the evolution of target-site resistance in weeds. Current knowledge on the biochemical mechanisms and the molecular basis behind target-site resistance in both diploid and polyploid weed species is summarized. Chapter 10 focuses on gene amplification as a target-site resistance mechanism in weeds and examines the applications of molecular cytogenetic tools to unravel some of the complex herbicide resistance mechanisms at chromosome level in weed species. Chapter 11 discusses the scope of molecular biology and genomics in weed science. It provides insights into the genomic approaches that can be applied to enhance our understanding of complex traits that enable weeds to adapt and survive in different environments.

Overall, this book is expected to be of particular interest to people working in both academia and the agro-based industry. This publication will also be important as a source of reference to the weed science/plant protection community. Most importantly, this book will be valuable for undergraduate and graduate students majoring in agronomy.

Contents

Chapter 1

Introduction

Advancement of Weed Science as an Important Discipline of Agriculture

Aruna Varanasi and *Mithila Jugulam**

What are Weeds and Why do We Care about Them?

About 10,000 years ago, mankind started encouraging the growth of desirable wild plants by cultivation, which marked the beginning of agriculture. Weeds (unwanted plants) came into existence with this beginning of agriculture and have become one of the major constraints for successful crop production. Mankind's effort to control weeds is an ongoing struggle that presents many challenges and difficulties. Weeds, therefore, play a major role in shaping the success of agricultural production, thus becoming an integral part of our domestic culture.

Weeds are not different from other plants that compete for survival. However, weeds are considered as plants growing 'out of place' or where they are not intentionally planted, and hence undesirable for human activities. Nature has its way of evolution of plant attributes for survival, and weeds being naturally resilient, tend to dominate and out compete the neighboring plants in an ecological niche. Typically, weeds possess one or more of the aggressive characteristics such as rapid establishment and growth, seed dormancy, vegetative reproductive structures, and abundant seed production. Weeds are also genetically diverse and highly adaptable which allow them to survive in diverse environments including sites disturbed by human activities, often interfering negatively with human activity.

Department of Agronomy, Kansas State University, Manhattan, KS.
* Corresponding author: mithila@ksu.edu

Weeds are considered as highly damaging agricultural pests. Worldwide, weeds cause extensive yield losses to crops and increase grower's production costs more than any other agriculture pests including insects, disease-causing pathogens, nematodes, birds, or rodents. Weeds are undesirable for agriculture for many reasons. Primarily they reduce crop yields by competing directly with crops for space, sunlight, water, and other essential soil nutrients. Other problems associated with weeds in crop production include reduced yield quality due to contamination, interference in harvest operations, allelopathy (release of natural substances that inhibit plant growth), and serving as hosts for crop pests or pathogens. Weeds are also a huge menace to rangelands, pastures, parks, and recreation areas. If left unchecked in these regions, they affect livestock production, reduce property values, create safety hazards, and destroy ecosystems. Efficient management of weeds is critical not only for boosting crop yields and sustaining crop production but also for maintaining the balance of ecosystems and preventing public health hazards.

Weed Control: Historical Perspectives and Current Scenario

Weed control is a major component of agriculture. In ancient times, humans paid little attention to weed control and most weeds were involuntarily uprooted or killed during tillage practices and seedbed preparation. Hand-weeding (pulling the weeds manually with hands) can be considered as the earliest known method of weed control. This was followed by use of some simple hand tools like a knife, hoe, or mattock which still required the employment of manual labor. These tools were used in conjunction with plow to break up the soil for planting which also helped in uprooting the weeds. With time, domestication of animals like oxen and horses enabled the use of larger tools for planting and other agricultural operations through animal traction. These methods continued for several centuries as manual labor, mainly in the form of women and children, was available in abundance. With the beginning of industrial age in the 18th century, a new era of weed control dawned in the form of invention and technology. Along with technological improvements in other areas of agriculture, there were some drastic changes in the methods used for weed control. Some of the earliest inventions of this age were made by a farmer in England, Jethro Tull who invented grain drills and cultivation tools (Timmons 2005). The subsequent widespread use of these tools made weed control much easier than before. The grain drills initiated the concept of row cropping which enabled easy identification of weeds between the rows. Subsequently, cultivation tools like horse-drawn harrows were used to control the weeds between the rows by uprooting and killing them. These cultivation tools were made more efficient and specialized by using them in conjunction with animal traction and tractors. A few examples of these specialized cultivation tools include flexible tine and rod weeders, wheel cultivators, straddle-row cultivators, riding cultivators, rotary hoes, spring tooth harrows, push-and-pull hoes, finger weeders, and French plow (mainly used for weed control in orchards). The use of such tools paved the way for mechanical weed control methods which remained the mainstay of weed control for the next three centuries.

In some parts of the world, cultural weed control methods like crop rotation and cover crops were used inadvertently to prevent weed emergence. Crop rotation and cover crops usage can be considered as two of the oldest agricultural practices that have not lost their significance even in this era. The purpose of using these practices in ancient times, when weed control received little attention, was to restore and maintain a productive soil. Farmers generally practiced three- or four-year rotation which included a legume crop and a fallow system to regenerate nutrients in the soil. These methods prevented the dominance of weeds specific to each crop in the rotation and thus helped in reducing the weed pressure unintentionally. On the other hand, crops with different growth habits were sown together to create cover crop effect and suppress weed growth. For example, in the US, three crops, corn with erect plant growth, squash with early germination and rapid growth, and beans with prostrate growth habit were sown together. The rapidly growing squash prevented weed emergence by creating a cover crop effect, whereas the upright growing corn provided support for the growth of beans.

With increase in world population and the subsequent demand for higher food production, agriculture entered into new era of intensive cropping programs and increased mechanization in the 19 and 20th centuries. Although this intensification resulted in increasing crop yields and reducing production costs, greater challenges regarding agricultural pests like weeds were encountered. More efficient and profitable methods of weed control were needed to overcome the yield losses caused by weeds. The advent of chemical weed control methods using herbicides was considered a boon to the farming community, especially in developed countries like the US, where availability of manual labor is limited. Herbicides were first used as chemicals for weed control intentionally in a crop in the mid-19th century. Previously, farmers were unable to comprehend the concept of chemicals being used selectively to remove weeds from crops as both weeds and crops are biologically similar organisms. The first herbicides were inorganic salts such as sodium chloride, sodium chlorate, arsenic salts and copper sulfate which were used for nonselective control of some weeds like common hawkweed, field bindweed, charlock, and wild mustard in the US (Call and Getty 1923; Jones and Orton 1896). Large amounts (@ 20 tons/acre) of sodium chloride were used for weed control along highways and railroad rights-of-ways in Kansas in the 1930s and 1940s (Yost 1940). Inorganic acids like sulfuric acid were also used for selective control of certain broadleaf weeds in cereals (Aslander 1927; Ball and French 1935). Other chemicals like iron sulfate, ammonium sulfate, gasoline, kerosene, and dinitros were used for selective control of broadleaf weeds in lawns, cereals, legumes, and vegetable crops (Barrons and Grigsby 1946; Crafts 1945; Harris and Hyslop 1942; Helgeson and Gebracht 1940; Litzenberger and Post 1943; Schenedimen et al. 1943). Many of these chemicals were used extensively in the Europe and the US, but eventually discontinued due to persistence in soil and concerns with environmental pollution. Although use of inorganic chemicals for weed control had varying degrees of success, they renewed research interest in chemical weed control and eventually led to the development of synthetic herbicides. The use of herbicides for weed management started gaining widespread recognition with the advent of synthetic compounds in the 1940s (Timmons 2005). Subsequently, the discovery of several new herbicide compounds opened up additional avenues for weed control and marked the real beginning of chemical weed control for improved crop production. The herbicide discovery period also saw some advances in herbicide application technology and other fundamental areas of weed science including biology/ecology of weeds and crop-weed competition.

The discovery of the first selective herbicides such as phenoxyacetic acids (e.g., 2, 4-D and MCPA) in the 1940s allowed for more consistent control of many broadleaf weeds in cereal crops. The weed killing properties of these herbicides quickly gained popularity and widespread use and revolutionized the cereal crop production. While phenoxyacetic acids provided reliable control of broadleaf weeds in corn, sorghum, small grains and other cereal crops, they had limited activity on grass weeds and also caused some crop injury. Soon after, new formulations like aliphatic acids and its derivatives and the carbanilates were developed which provided control of some weedy grasses. The development of other organic compounds such as acetamides, benzoic acids, acetic acids, substituted urea and uracil, phenylacetic acid derivatives, s-triazines and triazoles progressed simultaneously. Some of these herbicides had limitations in the form of marginal crop selectivity, shorter duration activity, narrow weed spectrum, and restricted activity in some soils and weather conditions. Nevertheless, they were commercialized for weed control and became available for some specific weed problems. By 1970s, over 120 organic herbicides were available in the market for use as herbicides in crops, industrial settings, lawns, and elsewhere (Timmons 2005). However, most of these herbicides had public health and safety concerns over residual activity in soil and environmental toxicity.

Under growing pressure, chemical companies focused on developing herbicides that could be used at lower rates but still had greater efficacy, lower soil persistence, and reduced toxicological effects on non-target species. Further discoveries led to the development of herbicides like glyphosate in the 1970s (Baird et al. 1971; Dill et al. 2010). In the beginning, glyphosate being a nonselective herbicide, its usage was limited to rights-of-ways, industrial yards, and for weed control after crop harvest and between rows in the orchards. However, with the discovery of genetically engineered herbicide-resistant technology in 1996, crops like soybean, corn, and cotton were made glyphosate-resistant which greatly revolutionized the methods used for weed control (Padgette et al. 1996). Growers embraced this technology rapidly as it provided a more reliable herbicide option in the form of glyphosate for postemergence broadspectrum weed control in such crops. The pairing of an herbicide with a resistant crop seed proved to be very effective and companies started investing and focusing on developing more number of crops with herbicide-resistant technology. Later, until early 21st century, several new compounds were developed, tested, and registered through various companies for weed control. Although no new herbicide modes of action have been discovered (Duke 2012) in the recent years, several new formulations for the existing modes of action are being tested to provide better options for weed management. For instance, new acetolactate synthase (ALS)-inhibitors such as mesosulfuron-methyl (Safferling 2005), thiencarbazone-methyl (Santel et al. 2012; Veness et al. 2008), and pyroxsulam (Wells 2008) have provided postemergence weed control options in crops like wheat, rye, corn, and other cereals. Although crop safeners are recommended with these herbicides to ensure crop selectivity, they are very effective in herbicide mixtures and have a significant market impact.

Another herbicide mode of action which has had some successful herbicide introductions in the recent years is the 4-hydroxyphenylpyruvate dioxygenase (HPPD)-inhibitors (Ahrens et al. 2013). The first generation HPPD products (pyrazolynate, pyrazoxyfen, benzofenap, and sulcotrione) were introduced in the 1980s and required relatively higher application rates for postemergence control of broadleaf weeds in crops like rice and corn (Ahrens et al. 2013; Beraud et al. 1993; Van Almsick 2012). However, the introduction of second generation HPPD-inhibitors like mesotrione and tembotrione in early 2000s, which belonged to the triketone family, significantly improved the market for HPPD products not only because of their potency at lower application rates but also because of their use for both premergence and postemergence applications (Edmund and Morris 2012). Since its introduction in 2001, mesotrione and its associated premixes have been widely used for control of broadleaf weeds in corn (Mitchell et al. 2001). On the other hand, tembotrione introduced in 2007, is being used in corn for postemergence control of many broadleaf and some grass weeds including weeds resistant to glyphosate, ALS-inhibitors, and dicamba (Van Almsick et al. 2009). Additional new representatives of the HPPD-inhibitors include postemergence herbicides such as pyrasulfatole (2008) for cereals, topramezone (2006) for corn, and tefuryltrione (2010) for rice (Kraehmer et al. 2014). Some of the newer herbicides, introduced in the late 2000s, from other modes of action include indaziflam (cellulose biosynthesis inhibitor), aminocyclopyrachlor (auxinic herbicide), saflufenacil (protoporphrinogen oxidase inhibitor), and pyroxasulfone (very long chain fatty acid inhibitor (Kraehmer et al. 2014)).

Evolution of Weed Science as a Discipline

Weed Science is a discipline of agriculture that focuses on integrating all aspects of weeds including the management of weed species in both agriculture and non-agriculture systems. Currently, the primary areas of weed science include biology, ecology, physiology, molecular biology and the genetics/genomics of weeds. Although the study of weeds has been a part of agricultural education for a long time, weed science emerged as a discipline and came into limelight upon introduction of herbicides for weed management in the 1950s. The discovery of synthetic organic herbicides and their subsequent commercialization provided a major boost to weed science and established it as a valid scientific discipline. Unlike entomology, plant breeding or plant pathology disciplines which have a richer historical lineage, weed science has a narrow history as most of its research was primarily directed towards improving weed management using chemical weed control methods. Herbicides represented a new technology providing greater benefits for the farming community as it boosted yields significantly with concomitant reduction in labor costs. This led to a growing and profitable market, and industries, university research and extension scientists diverted their efforts more on herbicides to develop them as tools to manage weed infestations and increase food production.

The rapid adoption of chemical weed control resulted in extensive and repeated use of herbicides in agriculture. Consequently, the evolution of resistance to herbicides has been documented in some weed species and has eventually led to significant shifts in the weed science research. The first case of herbicide resistance in weeds was reported in the 1950s (Hilton 1957) but at the time, herbicide resistance research received little attention as new herbicides were being introduced into the market which provided options for the management of herbicide-resistant weeds. However, the introduction of herbicide-resistant crops in the mid-1990s led to the extensive and exclusive use of a single mode of action of herbicide (e.g., glyphosate) which significantly increased the number of weed species that evolved resistance to herbicides. By mid-2000s, resistance to several herbicide modes of action including, EPSPS-, ALS-, ACCase-, and PSII-inhibitors was documented in all the economically important weeds of the US (Heap 2016). With no new herbicide modes of action in the horizon (Duke 2012), weed scientists and industry research specialists emphasized more on studying the mechanisms underlying herbicide resistance to understand the evolutionary pattern and spread of herbicide resistance. Around the same time, basic studies on weed biology and ecology, and weed management approaches that integrate cultural, mechanical, and chemical control methods received increasing attention. At a time when resistance issues result in the potential loss of certain herbicides, adoption of alternative weed control methods is critical to control weed populations as well as to ensure continued usage of certain herbicides. Weed scientists have also started an investigation of alternative measures, such as biological agents, mulches, cover crops, crop rotation, and crop planting density for the management of weeds. The aim of such research has been to develop best management practices that provide effective weed management and at the same time preserve the efficacy of important herbicides for future generations.

In light of all the recent developments in weed science research, this discipline has now developed into a more balanced and mature field and is considered as one of the most important areas of research in agriculture. Weed scientists will have more grueling tasks as new challenges develop in the form of evolution of complex mechanisms of resistance to herbicides (e.g., non-target site or target site based resistance), integrated weed management approaches, herbicide use in conservation systems and in sustaining herbicide efficacy in the context of global climate change. To address these issues, research efforts focused on weed population and evolutionary genetics, molecular cytogenetics, molecular biology, modeling, and weed biology have been intensified in the recent years.

Biology, Physiology, and Molecular Biology of Weeds

The aim of this book is to integrate three primary areas of weed science research; weed biology, physiology, and molecular biology. It is an attempt to reflect on what has been accomplished in these areas of weed science and also the direction of future research. The topics in each of these areas specifically emphasize on the current research scenario and the future perspectives in weed biology, physiology, and molecular biology.

Weed biology/ecology

Studies on the biology and ecology of weeds form a crucial part of weed science research as they play a major role in devising successful weed control methods. Any measure employed for weed control is aimed at shifting the balance of plant populations in an ecological community to the desired direction. Weed biological and ecological factors such as habitat, growth characteristics, competitive ability, geographic distribution, and prevalence determine the interrelationship of weed populations to the neighboring communities and the environment. Knowledge of weed seed germination, behavior and persistence in soil is necessary to understand the factors influencing weed population dynamics. Likewise, appreciation of the factors influencing weed reproduction, propagation, and seed dispersal is important to prevent the introduction and spread of weeds into new areas. This is especially crucial in the case of the most problematic weed species as well as the weed populations that have evolved resistance to herbicides. The chapters in this book cover new research findings in the areas of weed biology and ecology with emphasis on weed germination, demographics, weed seed bank dynamics, and how they influence herbicide resistance and weed management.

Weed physiology

The evolution of herbicide resistance in weeds and the increasing reports of herbicide resistance cases in several major weed species, worldwide, has significantly impacted the direction of weed research in the last 10 years. Weed scientists are now focusing on understanding the mechanisms underlying herbicide resistance in weed populations to predict the evolutionary trajectory of herbicide resistance. Resistance mechanisms vary with weed species, herbicide, and the environment, hence the need to be investigated on a case-by-case basis. This information is useful to design location-specific weed management strategies that delay further spread of resistance and ensure continuous use of herbicides as cost effective tools for weed management. Several studies have been conducted to investigate herbicide resistance mechanisms in different weed species using physiological techniques including target enzyme assays and radiolabeled herbicide bioassays that determine the absorption, translocation, and metabolism of herbicide parent molecules in the resistant plants. The Chapters 8 and 9 discuss some of the recent advances and future challenges in target and non-target herbicide resistance mechanisms. Information on the role of herbicide safeners and the physiological aspects involved in their mechanism of action for crop selectivity is also presented.

Molecular biology

Technological advances in molecular and genomic research including genetic engineering, DNA-based genotyping, and next generation sequencing has greatly facilitated rapid progress in crop improvement. The continuous decrease in the cost of molecular biology tools has now directed the weed science research to 'Genomics era'. Weed science research is increasingly becoming an interdisciplinary science as more number of weed scientists shift gears towards molecular biology and genomics in order to unravel some of the complex mysteries associated with weeds and their survival mechanisms. This book addresses some of the recent developments in weed molecular biology research. These developments include applications of cytogenetic and genomic tools for improved understanding of weed biology and herbicide resistance mechanisms. The Chapters 10 and 11 provide insights on current and prospective molecular approaches that facilitate the study of complex weedy traits and their role in weed adaptation and survival. Information generated through this research will not only support the existing weed control methods but also assist in devising new strategies for effective weed management.

Future directions

Weed science research is separated into two major categories: weed control research and weed principles research. While weed control research aims at studying weed control methods and their associated topics which include chemicals, tillage, biological controls and other studies, weed principles research focuses on more fundamental aspects of weed biology and ecology. As discussed above, until recently, much of the resources and research efforts have been directed towards the development and implementation of weed control methods, whereas weed principles research received little attention and thus we can assume that it is still in its infancy. It is essential to have a thorough understanding of the principles of weed science as it forms the core of all studies on weeds and lays the foundation for applied weed research that directly impacts the farming community. Therefore, it is important to enhance weed principles research to provide the much needed basic information necessary to understand weeds and the problems that they cause to human activities. Weeds are also important to study as model plants as they are genetically more diverse than crop plants. It is beneficial to study the traits that make the weeds more adaptable to adverse environmental conditions so as to apply the concepts to improve crops for sustainable crop production.

In developed countries like the US, weeds dictate crop production practices and the success of most cropping systems. Weeds are considered to be major deterrents for the development of successful crop production systems in current as well as in future climate conditions. Lack of effective weed control strategies is one of the major reasons for growers to not venture into alternative cropping systems. In order to combat this weed problem, new weed control technology must be based on a strong weed principles research. Weed scientists need to provide vision and support to other agricultural sciences like crop breeding, agronomy, crop physiology, crop ecology, and crop genetics to achieve the goal of developing new crop production systems that are more productive while being less destructive to the environment. To accomplish this, weed science research must seek active collaborations with scientists of other agricultural disciplines like plant breeding, horticulture, soil science, plant pathology, entomology, plant molecular biology, and genetics. Such collaborative research effort is bound to become successful in meeting the demands of food production in present and future conditions.

LITERATURE CITED

Ahrens H, Lange G, Müller T, Rosinger C, Willms L and van Almsick A (2013) 4-Hydroxyphenylpyruvate dioxygenase inhibitors in combination with safeners: Solutions for modern and sustainable agriculture. Angewandte Chemie International Edition 52: 9388–9398.

Aslander A (1927) Sulfuric acid as a weed spray. J. Agric. Res. 34: 1065–1091.

Baird DD, Upchurch RP, Homesley WB and Franz JE (1971) Introduction of a new broadspectrum postemergence herbicide class with utility for herbaceous perennial weed control. pp. 64–68. *In*: Proc North Cent Weed Control Conf.

Ball WE and French OC (1935) Sulfuric acid for control of weeds. California Agric. Exp. Sta. Bull. 596: 29 p.

Barrons KC and Grigsby BH (1946) The control of weeds in canning peas. Michigan Agric. Exp. Sta. Quart. Bull. 28: 145–156.

Beraud M, Claument J and Montury A (1993) ICIA 0051, a new herbicide for control of annual weeds in maize. Proc. Br. Crop. Prot. Conf. Weeds 51–56.

Call IE and Getty RE (1923) The eradication of bindweed. Kansas Agric. Exp. Sta. Circ. 101: 18 p.

Crafts AS (1945) A new herbicide, 2, 4-dinitro-6-secondary butylphenol. Science 101: 417–418.

Dill GM, Sammons RD, Feng PC, Kohn F, Kretzmer K, Mehrsheikh A, Bleeke M, Honegger JL, Farmer D, Wright D and Haupfear EA (2010) Glyphosate: discovery, development, applications, and properties. pp. 1–33. *In*: Nandula VK (ed.). Glyphosate Resistance in Crops and Weeds: History, Development, and Management. John Wiley and Sons, Inc., Hoboken.

Duke SO (2012) Why have no new herbicide modes of action appeared in recent years? Pest Manag. Sci. 68: 505–512.

Edmunds AJF and Morris JA (2012) Hydroxyphenylpyruvate dioxygenase (HPPD) inhibitors: triketones. pp. 249–254. *In*: Krämer W, Schirmer U, Jeschke P and Witschel M (eds.). Modern Crop Protection Compounds. Wiley-VCH, Weinheim, Germany.

Harris IE and Hyslop DJR (1942) Selective sprays for weed control in crops. Oregon Agric. Exp. Sta. Bull. 403: 31 p.

Heap I (2016) The International Survey of Herbicide Resistant Weeds. http://www.weedscience.org/summary/home.aspx (October 16, 2016).

Helgeson EA and Gebracht DD (1940) Chemical control of annual weeds in flax and grain fields. North Dakota Agric. Exp. Sta. Bimonthly Bull. 3: 7–10.

Hilton HW (1957) Herbicide tolerant strains of weeds. Hawaiian Sugar Planters Association Annual Report. 69 p.

Jones IR and Orton WG (1896) The orange hawkweed or "paint-brush." Vermont Agric. Exp. Sta. Bull. 56: 15 p.

Kraehmer H, van Almsick A, Beffa R, Dietrich H, Eckes P, Hacker E, Hain R, Strek HJ, Stuebler H and Willms L (2014) Herbicides as weed control agents: state of the art: II. Recent achievements. Plant Physiol. 166: 1132–1148.

Litzenberger SC and Post AH (1943) Selective sprays for control of weeds in Kentucky bluegrass lawns. Montana Agric. Exp. Sta. Bull. 411: 23 p.

Mitchell G, Bartlett DW, Fraser TE, Hawkes TR, Holt DC, Townson JK and Wichert RA (2001) Mesotrione: a new selective herbicide for use in maize. Pest Manag. Sci. 57: 120–128.

Padgette SR, Re DB, Barry GF, Eichholtz DE, Delannay X, Fuchs RL, Kishore GM and Fraley RT (1996) New weed control opportunities: development of soybeans with a roundup ready gene. pp. 53–84. *In*: Duke S (ed.). Herbicide Resistant Crops: Agricultural, Economic, Environmental, Regulatory, and Technological Aspects. CRC Press, Boca Raton, FL.

Safferling M (2005) Atlantis. Pflanzenschutz-Nachr Bayer 58: 165–299.

Santel HJ (2012) Thiencarbazone-methyl (TCM) and Cyprosulfamide (CSA)–a new herbicide and a new safener for use in corn. Julius-Kühn-Archiv 434: 499.

Schinendiman A, Torrie JH and Briggs GM (1943) Effects of sinox, a selective weed spray, on legume seedlings, weeds, and crop yields. J. Amer. Soc. Agron. 35: 901–908.

Timmons FL (2005) A history of weed control in the United States and Canada 1. Weed Sci. 53: 748–761.

Van Almsick A, Benet-Buchholz J, Olenik B and Willms L (2009) Tembotrione, a new exceptionally safe cross-spectrum herbicide for corn production. Pflanzenschutz-Nachr Bayer 62: 5–16.

Van Almsick A (2012) Hydroxyphenylpyruvate dioxygenase (HPPD) inhibitors: heterocycles. pp. 262–274. *In*: Krämer W, Schirmer U, Jeschke P and Witschel M (eds.). Modern Crop Protection Compounds. Wiley-VCH, Weinheim, Germany.

Veness J, Patzer KTA and Steckler MK (2008) Thiencarbazone-methyl, a new herbicide active ingredient in Canada. Canadian Weed Science Society 62nd Annual Meeting. CWSS-SCM, Pinawa, Canada. 34 p.

Wells G (2008) Pyroxsulam for broad-spectrum weed control in wheat. pp. 297–299. *In*: Van Klinken RD, Osten VA, Panetta FD and Scanlan JC (eds.). Proc. Weed Management, 16th Australian Weeds Conf. Queensland Weeds Society, Brisbane, Australia.

Yost TF (1940) Summary of bindweed situation and progress in 1939. Kansas State Board Agric. Rep. 237: 73 p.

Chapter 2

Targeting Weed Seedbanks
Implications for Weed Management

Prashant Jha,[1,]* *Rakesh K. Godara*[2] *and Amit J. Jhala*[3]

INTRODUCTION

Weed seedbanks are dynamic in nature, both temporally and spatially (Cardina et al. 1996, 1997; Colbach et al. 2000). Cavers (1995) stated that "a weed seedbank serves as a genetic memory in the soil" for the weed community, and represents both the past and future weed invasions in an agroecosystem (Buhler et al. 1997a; Cardina et al. 2002; Swanton and Booth 2004). The seedbank characteristics influence the population dynamics of both annual and perennial weed species and the success of weed management practices (Buhler et al. 1997a, b). For annual weeds, seeds are the only source of population increase, and seed dormancy is the primary means of persistence of those species in the soil seedbank (Baskin and Baskin 1998). Weed infestation in a given field is largely influenced by the number, density, and distribution of each species across the field, and these factors further dictate the efficacy of weed control programs (Buhler et al. 1997a, b). Greater efforts are needed to control weeds such as Palmer amaranth (*Amaranthus palmeri* S. Wats.) and kochia [*Kochia scoparia* (L.) Schrad.], both of which input higher number of seeds into the soil seedbank (Taylor and Hartzler 2000; Davis 2006; Kumar and Jha 2015a, b).

[1] Assistant Professor of Weed Science, Montana State University-Bozeman, Southern Agricultural Research Center, Huntley, MT 59037, USA.
[2] Research Scientist, Monsanto Company, St. Louis, MO, USA.
[3] Assistant Professor of Weed Science, Department of Agronomy and Horticulture, University of Nebraska–Lincoln, Lincoln, NE 68583, USA.
* Corresponding author: pjha@montana.edu

Mathematical models have been used to determine the potential of weed species to contribute to weed seedbank. For instance, Davis (2006) adopted a modelling approach to target annual weed seedbanks using giant foxtail (*Setaria faberi* Herrm.) and common lambsquarters (*Chenopodium album* L.) as model weed species. For giant foxtail (a weed species with modest fecundity and low seed dormancy) control methods aimed at increasing the mortality of dormant weed seed bank were more effective in reducing the population growth rate compared with the methods that targeted the above ground life stages. In contrast, for common lambsquarters (a species exhibiting high fecundity and high seed dormancy) weed control tactics aimed at reducing the seed dormancy or seedbank persistence were just as important as those for controlling the above ground life stages (Davis 2006). Therefore, in addition to effective seedling-focused tactics (herbicides and tillage/cultivation), strategies to reduce the weed seedbank inputs and seedbank persistence should be an important component of long-term weed management programs.

The concept of a "critical period of weed control (CPWC) or critical weed-free period (CWFP)" is mainly focused on limiting crop yield loss (Knezevic et al. 2002), with less emphasis on weed seed inputs to the soil. Several studies have documented that early-season herbicide escapes and late-emerging weed cohorts, although having little to no impact on crop yields, can contribute significant amounts of seeds to replenish the soil seedbank (Jha et al. 2008; Wilson and Sbatella 2011; Kumar and Jha 2015b), facilitating future weed problems in agricultural fields. Instead of solely considering annual crop yield loss by killing weed seedlings, multiple weed control tactics should be utilized to increase seed mortality and deplete the weed seedbank over multiple growing seasons (Davis and Liebman 2003; Gallant 2006). A more aggressive method of weed seedbank management is "no-seed threshold strategy" proposed by Jones and Medd (2000) and Norris (1999). It has been proven that seed survival has a greater influence on weed population dynamics compared with above ground plant survival (Jordan et al. 1995; Swanton and Booth 2004).

A framework for weed seedbank management proposed by Swanton and Booth (2004) is comprised of direct and indirect methods of control targeted at multiple weed life history stages depending on weed biology, density, seed bank persistence, cropping system, and available weed control tools. The direct methods include increasing seed mortality and manipulating seed dormancy and seedling emergence, while indirect methods include eliminating aboveground weed biomass and reducing seed production. Management practices have a strong impact on the weed seed life cycle and represent opportunities for regulating seed bank characteristics such as size, species composition, persistence, and distribution (Schweizer and Zimdahl 1984; Burnside et al. 1986).

Weed Seedbank Management Strategies

Herbicides

Herbicide-based weed management systems generally have a higher weed control efficacy compared to mechanical or cultural weed control systems, and consequently have fewer seed bank additions (Gallant 2006). There is an enormous amount of literature available on herbicide efficacy and control of emerged weed seedlings; however, very few studies have focused on the long-term impacts of herbicides on weed seedbanks. Impact on weed seedbanks from herbicide escapes or late-season cohorts that emerge after herbicide activity have not been evaluated routinely (Bagavathiannan and Norsworthy 2012). In general, a linear or polynomial decrease in seed production is observed with an increase in herbicide rate (Roggenkamp et al. 2000; Blackshaw et al. 2006), and herbicide responses are influenced by the levels of crop-weed competition for available resources, crop management practices, and herbicide-weed species complex (Blackshaw et al. 2006; Nurse et al. 2008). For instance, use of herbicide rates lower than the recommended rates may translate into higher weed seed production; however, integrating other management tactics such as increased crop seeding rates or cultivation may reduce the risks associated with an increased weed seed rain at lower than recommended herbicide rates, and may be as effective as a recommended-rate herbicide program for managing the weed seedbank (Malugeta and Stoltenberg 1997; O'Donovan et al. 2004).

In a three year study on the effect of a single POST application of glyphosate/glufosinate at 0.5x, 0.75x, or 1x rates (where x is the recommended full use-rate of the herbicide) on the weed seedbank density in continuous corn (*Zea mays* L.) or corn/soybean [*Glycine max* (L.) Merr.] rotations, the weed seedbank decreased with an increase in herbicide rate after three years; nevertheless, the lowest seedbank treatment (corn/soybean rotation with a 1x rate of glyphosate every year) still had higher numbers of monocot and dicot weed seeds per m² compared with the initial seedbank (Simard et al. 2011). This emphasizes the need for managing the weed seedbanks with multiple control tactics rather than a single herbicide application per season.

Multiple POST herbicide applications are often needed to prevent seed production from weed cohorts that emerge late in the season (Jha et al. 2008; Kumar and Jha 2015a, b). Pusley species that escaped a single application of glyphosate at the V3 or V6 stage of glyphosate-resistant (GR) soybean, produced up to 15,870 seeds per m² compared with no seed production with three sequential applications of glyphosate at the V3, V6, and R2 stages of GR soybean (Jha et al. 2008). Palmer amaranth seedlings that emerged after the V3 stage of application in glyphosate-resistant soybean, produced 600 seeds per m², and two to three applications at the V3, V6 or V3, V6, R2 stages of soybean were needed to prevent seedbank replenishment from late-emerging cohorts (Jha et al. 2008).

Use of PRE soil-residual herbicides can not only reduce the risks of early-season weed competition and crop yield losses, but can also significantly reduce the population density of the emerged weed seedbank (Wilson et al. 2011; Norsworthy et al. 2012). For instance, addition of PRE herbicides flumeturon plus prometryn and pendimethalin to the glyphosate POST program reduced barnyardgrass [*Echinochloa crus-galli* (L.) Beauv.] emergence to 30 plants per m^2, compared with 285 and 176 plants per m^2 in the glyphosate only and glyphosate, plus a grass herbicide POST program in glyphosate-resistant cotton (*Gossypium hirsutum* L.) (Werth et al. 2008). The researchers reported a hundred fold reduction in the weed seedbank in the presence of soil-residual herbicides compared to 10–20 fold reductions in the treatments with no soil-residual herbicides. In a study about the effect of weed management on seedbank dynamics of GR vs. glyphosate-susceptible (GS) horseweed in Indiana, Davis et al. (2009) reported that a PRE soil-residual herbicide followed by a non-glyphosate POST herbicide resulted in a noticeable shift in the weed seedbank from an initial GR:GS ratio of three to one to a ratio of one to six after four years.

Crop topping, which is a POST application of non-selective herbicides before crop harvest to prevent weed seedset is a valuable tool for reducing seedbank inputs (Taylor and Oliver 1997; Walsh and Powles 2007). A late-season (early inflorescence stage) application of glufosinate, 2, 4-D, or dicamba reduced seed production of Palmer amaranth by 87 to 95 percent compared to the nontreated control (Jha and Norsworthy 2012). Furthermore, glufosinate, 2,4-D, and glyphosate reduced seed weight by 22 percent, and seed viability by 45 to 61 percent compared with 97 percent viability of seeds produced by the nontreated plants (Jha and Norsworthy 2012). Glyphosate applied at anthesis prevented viable seed production in wild oat (*Avena fatua* L.) (Shuma et al. 1995). Herbicides applied postharvest prevented the addition of 1,200 seed per m^2 (12 million seed per ha) of glyphosate-resistant Palmer amaranth to the soil seedbank (Crow et al. 2015). Late-season or postharvest herbicide applications are especially needed to prevent infestations of weed species such as kochia and Russian thistle, which can regrow after crop harvest and produce significant amount of seeds (Young 1986; Schillinger and Young 2000; Kumar and Jha 2015a). Geographic information system (GIS)-based weed maps and use of light-activated, sensor-controlled spray technology provide new opportunities for postharvest weed control, and could potentially reduce herbicide costs as well as increase grower profitability and environmental sustainability (Riar et al. 2011). From an herbicide resistance (HR) evolution standpoint, late-season weed seed production, if not prevented, can increase the probability of occurrence of HR weed seedbank and perpetuation of rare resistant individuals (Jasieniuk et al. 1996; Bagavathiannan and Norsworthy 2012). This may be more important for agricultural systems with high risks of HR evolution and for weed species with extended emergence periods and prolific seed production, such as Palmer amaranth, common waterhemp (*Amaranthus rudis* Sauer), and kochia (Norsworthy et al. 2012; Kumar and Jha 2015a, b).

Crop rotation

The effects of crop rotation on weed seedbanks are mediated by crop sequences that employ varying levels of planting and harvest dates, resource competition, fertility requirements, soil disturbance regimes, crop growth habits, and agronomic or weed management practices (Ball 1992; Buhler et al. 1997a; Cardina et al. 2002). Weeds that survive and produce seeds in one crop contribute to the seed bank establishment and infestations in successive crops. Diversified crop rotations provide greater opportunities for managing weed seedbanks over time and space than monoculture (Cardina et al. 2002). Weed community diversity tends to be greater in crop rotations than in monoculture (Ball 1992; Cardina et al. 2002), and is attributed to the variation in weed control practices, mainly herbicide use patterns during crop rotations, which tends to disrupt the life cycle of a more specific, well-adapted weed species, thereby creating niches for a greater diversity in the weed spectrum (Forcella et al. 1993; O'Donovan et al. 2007).

Cardina et al. (2002) proposed that crop rotation acts as an important environmental filter for determining the relative abundance and species composition of the weed seedbanks. Relative importance (RI) values based on relative density and relative frequency of each species is a measure of weed species shift (Cardina et al. 2002). In a 35 year crop rotation study, RI values for common lambsquarters, redroot pigweed (*Amaranthus retroflexus* L.), nimblewill (*Muhlenbergia schreberi* J.F. Gmel.), yellow foxtail [*Setaria pumila* (Poir.) Roemer and J.A. Schulters], purple deadnettle (*Lamium purpureum* L.), and yellow nutsedge (*Cyperus esculentus* L.) were lower in a corn–soybean rotation than in a continuous corn. Similarly, RI of grassy weeds like giant foxtail, was lower in corn–oat–hay rotation than in continuous corn (Cardina et al. 2002). Schreiber (1992) further reported that giant foxtail density was significantly reduced in soybean–corn and soybean–wheat–corn rotations compared to corn monoculture. The use of competitive crops in rotation provides opportunities to reduce weed competition and seed inputs to the soil seedbank, and can also benefit growers by reducing management inputs and costs (Kegode et al. 1999).

Herbicide use in different crops in rotation has a strong influence on the abundance and adaptability of weed species present in the soil seedbank (Vencill et al. 2012). Grass weeds can be more challenging to control in a grass crop, but can be more effectively controlled by herbicides in a broadleaf crop grown in rotation with the grass crop. Diversifying crop rotation is a key to long-term management of HR weed seedbank. A more diverse crop rotation allows the use of multiple herbicides having different sites of action to avoid recurrent selection pressure from the use of a single site-of-action herbicide (Norsworthy et al. 2012; Vencill et al. 2012). Wild oat resistance to triallate herbicide occurred after 18 years of annual triallate use in continuous wheat, but not in a wheat-fallow rotation, where the herbicide was applied 10 times over the same 18 year period (Beckie and Jana 2000). This implies that a more diversified rotation slows the evolutions of herbicide resistance in weeds and also provides opportunities for eliminating the rare resistance alleles from the soil seedbank (Neve et al. 2009; Norsworthy et al. 2012). Similarly, resistance to ACCase- and ALS-inhibitors in wild oat populations from Alberta, Canada were attributed to lack of diversity in crop rotation, e.g., no fall-seeded or perennial crops in the spring cereal monoculture (Beckie et al. 2004).

Recently, simulation models have been developed to predict the risk of evolution of glyphosate resistance in weed species in glyphosate-resistant cropping systems (Neve et al. 2009, 2011; Bagavathiannan et al. 2013). One such model predicted the evolution of glyphosate-resistant barnyardgrass within 9 year of continuous glyphosate use in glyphosate-resistant cotton monoculture in the southern U.S., and predicted the risk of occurrence would be as high as 47 percent in 15 years. However, the risk of resistance evolution could be reduced by 25 percent and almost six years, if glyphosate-resistant cotton was rotated with glyphosate-resistant or glufosinate-resistant (Liberty-Link™) corn (Bagavathiannan et al. 2013). Due to a greater diversity in type and timing of herbicides, the glyphosate-resistant soybean–corn rotation had 2.5- and 3-fold decrease in glyphosate-resistant horseweed seedbank densities than the continuous soybean (Davis et al. 2009). With limited herbicide options for controlling glyphosate-resistant kochia in GR sugar beet (*Beta vulgaris* subsp. vulgaris) in the northern U.S. Great Plains, managing the resistant weed seedbank with effective, non-glyphosate herbicides labeled in rotational crops such as barley (*Hordeum vulgare* L.), corn, or dry bean (*Phaseolus vulgaris* L.) becomes imperative (Kumar and Jha 2015a).

Tillage

Tillage influences the vertical distribution of weed seeds in the soil profile (Ball 1992; Clements et al. 1996; Cardina et al. 2002). In a study conducted by Yenish et al. (1992) in no-till systems with minimum soil disturbance, most weed seeds remained on or near the soil surface, with over 60 percent in the top one cm of the soil profile and only a few seeds remained below the top five cm in the soil. In a chisel plow tillage system, over 30 percent of weed seeds were found in the top one cm of the soil profile and the concentration of those seeds decreased linearly with increasing soil depth. Moldboard plowing, the most intensive tillage system, had fewer seeds in the upper 20 cm of the soil profile compared to chisel plow and no-till systems (Yenish et al. 1992). Also, Pareja et al. (1985) found 85 percent of all weed seeds concentrated in the upper 5 cm soil profile in a reduced tillage system compared with only 28 percent in the same depth in a moldboard plow system. Similarly, the top five cm of the soil profile contained only 37 percent of the weed seedbank in moldboard plow compared with 61 and 74 percent in chisel plow and no-till system, respectively (Clements et al. 1996). However, moldboard plow results in the most uniform distribution of weed seeds over the 20-cm soil depth (Yenish et al. 1992; Clements et al. 1996).

Cardina et al. (1991) reported a decline in weed species diversity with an increase in soil disturbance. For instance, after 25 years of continuous no-tillage, minimum tillage, or conventional tillage corn production, the number of weed species in no-tillage plots was greater than that of minimum or conventional tillage plots. Mean weed seedling emergence depths are smallest in no-tillage, followed by minimum tillage, and then conventional tillage plots (Cardina et al. 1991). However, in a study conducted by Steckel et al. (2007), the number of common waterhemp seeds remaining in the top zero to six cm of the soil after four years of seed rain, was greater in conventional-tilled plots compared to no-tilled plots due to greater germination and the emergence of common waterhemp in no-tillage treatments; hence, tillage increased the persistence of common waterhemp in the soil seedbank. Burial through tillage enhanced seed dormancy of wild oat, resulting in increased persistence of wild oat in the seedbank (Jana and Thai 1987). For several other annual weed species, the depth of burial through tillage plays a significant role in the persistence of the weed seedbank, with a smaller proportion of the seedbank remaining viable at depths less than five cm compared with burial depths of 10 cm or more (Roberts and Feast 1972). Therefore, tillage practices that bury seeds deeper into the soil may lead to build-up of the seedbank, resulting in long-term weed problems. Conversely, seeds lying on the soil surface in no-till systems are more likely to be depleted through germination, desiccation, predation, or microbial decay (Clements et al. 1996; Liebman and Davis 2000).

The effect of tillage on weed seedbanks in production fields is a complex phenomenon. Although persistence of small-seeded weed species will be relatively less in no-till than conventional-till systems (Clements et al. 1996; Steckel et al. 2007), the more rapid seed bank turnover in no-till vs. conventional-till may expose a greater proportion of the seedbank to herbicides. This interaction with increased herbicide use may result in an increased risk of selection of rare resistance alleles in the weed population (Beckie et al. 2008). In a survey conducted on risk assessment of herbicide resistance, 21 percent of the crop area with intensive tillage did not receive any herbicide application, whereas 99 percent of the crop area with no-tillage received at least one herbicide application, suggesting that the reduced-till systems are more reliant on herbicides than conventional tilled systems. Consequently, the occurrence of weed resistance was higher in no-tillage than minimum tillage or conventional tillage systems (Beckie et al. 2008).

Deep tillage using a moldboard plow is a viable strategy for reducing germination and emergence of small-seeded weed species. For example, deep tillage alone reduced Palmer amaranth emergence by 81 percent in soybean (DeVore et al. 2013). Common waterhemp emergence was reduced by 73 to 98 percent with deep tillage alone in the absence of a crop (Leon and Owen 2006). It has also been proposed that some level of tillage could potentially reduce the risk of herbicide resistance evolution by reducing the selection pressure from herbicide use or slowing the rate of weed seedbank turnover (Beckie et al. 2008; Price et al. 2011).

Another method of tillage often used to reduce weed seedbanks is the stale seedbed technique. In this approach, the soil is tilled at regular intervals for two to four wk before crop planting to stimulate weed seed germination. Irrigation is sometimes used to trigger germination and the emerged weeds are then killed by herbicides in conventional systems or by mowing or flaming in organic systems, and the crop is then planted (Johnson III and Mullinix 1995, 1998). This technique is more effective for small-seeded weed species such as common lambsquarters and pigweeds, which are buried in the soil, and require exposure to light and temperature fluctuations for seed germination (Baskin and Baskin 1998). Stale seedbed techniques have been used in a number of different crops, including corn, soybean, cotton, rice, peanuts, and vegetables and have been proven to be effective in reducing weed seedbank (Gunsolus 1990; Johnson III and Mullinix 1995, 1998; Shaw 1996).

Cover crops

The impact of cover crops on weeds can either be in the form of living plants or as plant residue after the cover crop is terminated (Teasdale et al. 2007). Living cover crops can have a greater suppressive effect on all phases of a weed's life cycle compared with the cover crop residues. A living cover crop can reduce light availability and red far-red ratio, and inhibit phytochrome-mediated seed germination of weeds such as *Amaranthus* species (Teasdale and Daughtry 1993; Gallagher and Cardina 1998; Jha et al. 2010). A living cover crop can provide direct early-season competition for light and nutrients and reduce weed emergence, growth, and seed production (Brainard and Bellinder 2004; Reddy and Koger 2004; Teasdale et al. 2007), and enhance weed seed predation at the soil surface (Davis and Liebman 2003). As high as 97 percent reduction in weed biomass has been reported with a live cover crop of rye (*Secale cereale* L.), mustards (*Brassica* spp.), barley (*Hordeum vulgare* L.), oat (*Avena sativa* L.), sweetclover [*Melilotus officinalis* (L.) Lam.], or hairy vetch (*Vicia villosa* Roth) (Teasdale and Daughtry 1993; Blackshaw et al. 2001; Grimmer and Masiunas 2004; Reddy and Koger 2004). In studies conducted in Georgia and South Carolina, USA, wild radish (*Raphanus raphanistrum* L.) and rye cover crops reduced total weed density up to 50 percent at 4 wk after planting sweet corn, in the absence of herbicides (Malik et al. 2008).

Cover crop residues on the soil surface can also reduce weed seed germination by reducing the available light and daily soil temperature amplitude (Teasdale and Mohler 1993), and suppress weed seedling growth by acting as a physical impediment (Teasdale et al. 2007). When cover crops are terminated or incorporated into the soil, the decomposition process can release phytotoxins (allelochemicals) or pathogens that inhibit germination and early-season growth of weeds (Davis and Liebman 2003; Norsworthy et al. 2007; Malik et al. 2008). In a research conducted by Teasdale et al. (1991), residues of hairy vetch and cereal rye (*Secale cereale* L.) reduced total weed density by 78 percent compared with plots with no cover crop. The allelochemicals BOA [(3H)-benzoxazolinone] and DIBOA [(2,4-dihydroxy-1,4-(2H) benzoxazine-3-one] released from rye after termination and incorporation into the soil inhibited emergence and seedling growth of Palmer amaranth, large crabgrass [*Digitaria sanguinalis* (L.) Scop.], barnyardgrass, and goosegrass [*Eleusine indica* (L.) Gaerth] (Burgos and Talbert 2000). Ten different glucosinolates, which are allelopathic compounds, were released by wild radish residues, which inhibited germination, emergence, and growth of prickly sida (*Sida spinosa* L.), pitted morningglory (*Ipomoea lacunosa* L.), sicklepod [*Senna obtusifolia* (L.) Irwin & Barneby], ivyleaf morning glory (*Ipomoea hederacea* Jacq.), and large crabgrass [*Digitaria sanguinalis* (L.) Scop.] (Norsworthy 2003; Malik et al. 2008). Rye, wild radish, or hairy vetch cover crop with half the recommended rates of atrazine plus S-metolachlor application resulted in season-long weed control and prevented sweet corn yield reductions (Burgos and Talbert 1996; Malik et al. 2008). Fall-seeded crimson clover and hairy vetch provided 58 to 62 percent control of glyphosate-resistant Palmer amaranth, 14 d before POST herbicide applications in corn. Additionally, corn was taller at V5 and V7 stages following a hairy vetch cover crop (Wiggins et al. 2015). Researchers concluded that although cover crops are not a stand-alone weed control method, integration of cover crops with herbicide mixtures (multiple modes of action) aids in reducing glyphosate selection pressure for the evolution of resistance in conservation-tilled, glyphosate-resistant corn and cotton

(Norsworthy et al. 2011; Price et al. 2011; Wiggins et al. 2015). Thus, cover crops serve as an additional tool for managing weed seedbanks.

Harvest weed seed control

The widespread occurrence of glyphosate-resistant and other multiple herbicide-resistant weeds in global agricultural production systems has resulted in a shift in weed control paradigm, with the development of alternative, non-chemical weed control technologies such as harvest weed seed control (HWSC), aimed at weed seed retention (capture) and destruction at crop harvest (Walsh et al. 2013). This technology offers opportunity for minimizing fresh seed inputs to the soil seedbank (Walsh et al. 2013). Harvest weed seed control systems have been successfully implemented in Western Australian cereal cropping systems for managing multiple herbicide-resistant annual ryegrass and wild mustard seedbanks (Walsh and Newman 2007; Walsh and Powles 2007). More commonly used HWSC systems include chaff carts, narrow windrow burning, bale direct, and Harrington Seed Destructor (HSD) (Walsh et al. 2013). Chaff cart collectors have been shown to remove 56 to 80 percent of the annual ryegrass seeds at grain harvest (Walsh and Powles 2007). Narrow windrow (500 to 600 mm) burning of chaff and residues collected after grain harvest was effective in killing 99 percent of seeds produced by annual ryegrass and wild radish (Walsh and Newman 2007). Bale direct system after harvest has been shown to remove 95 percent of weed seeds produced in the field (Walsh and Powles 2007) and HSD system has been shown to be effective in destroying 95 to 99 percent of weed seeds in the chaff collected at crop harvest (Walsh et al. 2012). The total weed seed retention at a harvest cutting height above 15 cm in a wheat crop was 85, 99, 77, and 84 percent for annual ryegrass, wild radish, brome grass, and wild oat, respectively (Walsh and Powles 2014). High weed seed retention of annual weeds at crop maturity and destruction of weed seeds at or after crop harvest, in combination with herbicide use, are effective in achieving targeted long-term low weed seedbank densities and managing herbicide resistance (Walsh et al. 2013; Walsh and Powles 2014).

Weed seed predation

Weed seed consumption and destruction by insects and mammals, known as seed predation, is a biological control method for management of weed seed bank. There are two types of dispersals. Weed seed predation occur when seeds are on the plant known as pre-weed dispersal, while weed seed predation after the weed seeds have dispersed and fallen from the mother plant known as post-weed seed dispersal. Post weed seed dispersal is the most common and birds, insects, and rodents are the most important predators. Scientific literature is limited about effects of predators on weed seed predation. Brust and House (1988) reported that rodents were important predators in no-tillage systems. Cromer et al. (1999) reported that mice consumed 10 to 20 percent seeds of common lambsquarters and barnyardgrass in no-till corn and soybean fields in Ontario, Canada. White et al. (2007) conducted greenhouse and field studies about feeding preferences of invertebrate seed predators and their effect on weed emergence. They conclude that feeding choice may vary by weed, predator species, and by seed burial depth; and it may reduce the weed emergence. More research is needed to fully explore the use of predators for weed seed predation at the level that reduce weed seedbanks.

Conclusion

A better understanding of the weed seedbanks is necessary for the long-term success of weed management systems. With escalating cases of herbicide resistance globally, there is a need to bring a shift in weed control paradigm, with more efforts aimed at managing the weed seedbanks over an extended time frame rather than simply controlling weed seedlings in production fields on an annual basis. The knowledge of various factors influencing weed seedbank dynamics will provide the foundation for development of innovative, cost effective, and sustainable weed control technologies. An integration of different practices to target and manage weed seedbanks as discussed in this chapter will aid in the long-term management of HR weed seedbanks.

LITERATURE CITED

Ball DA (1992) Weed seedbank response to tillage, herbicides, and crop rotation sequence. Weed Sci. 40: 654–659.

Baskin CC and Baskin JM (1998) Seeds: Ecology, Biogeography, and Evolution of Dormancy and Germination. Academic Press, San Diego.

Bagavathiannan MV, Norsworthy JK, Smith KL and Neve P (2013) Modeling the evolution of glyphosate resistance in barnyardgrass (*Echinochloa crus-galli*) in cotton-based production systems of the midsouthern United States. Weed Technol. 27: 475–487.

Bagavathiannan MV and Norsworthy JK (2012) Late-season seed production in arable weed communities: management implications. Weed Sci. 60: 325–334.

Beckie HJ and Jana S (2000) Selecting for triallate resistance in wild oat. Can. J. Plant Sci. 80: 665–667.

Beckie HJ, Hall LM, Meers S, Laslo JL and Stevenson FC (2004) Management practices influencing herbicide resistance in wild oat. Weed Technol. 18: 853–859.

Beckie HJ, Leeson JY, Thomas AG, Hall LM and Brenzil CA (2008) Risk assessment of weed resistance in the Canadian prairies. Weed Technol. 22: 741–746.

Brainard DC and Bellinder RR (2004) Weed suppression in a broccoli-winter rye intercropping system. Weed Sci. 52: 281–290.

Brust GE and House GJ (1988) Weed seed destruction by arthropods and rodents in low input soybean agroecosystems. Am. J. Alter. Agric. 3: 19–25.

Blackshaw RE, Larney FJ, Lindwall CW, Watson PR and Derksen DA (2001) Tillage intensity and crop rotation affect weed community dynamics in a winter wheat cropping system. Can. J. Plant Sci. 81: 805–813.

Blackshaw RE, O'Ddonovan JT, Harker K, Clayton GW and Stougaard RN (2006) Reduced herbicide doses in field crops: a review. Weed Biol. Manage. 6: 10–17.

Burgos NR and Talbert RE (1996) Weed control and sweet corn (*Zea mays* var. rugosa) response in no-till system with cover crops. Weed Sci. 44: 355–361.

Burgos NR and Talbert RE (2000) Differential activity of allelochemicals from *Secale cereale* in seedling bioassays. Weed Sci. 48: 302–310.

Buhler DD, Hartzler RG and Forcella F (1997a) Implications of weed seedbank dynamics to weed management. Weed Sci. 45: 329–336.

Buhler DD, Hartzler RG and Forcella F (1997b) Weed seed bank dynamics. J. Crop Prod. 1: 145–168.

Burnside OC, Moomaw RS, Roeth FW, Wicks GA and Wilson RG (1986) Weed seed demise in soil in weed-free corn (*Zea mays*) production across Nebraska. Weed Sci. 34: 248–251.

Cardina J, Regnier E and Harrison K (1991) Long-term tillage effects on seed banks in three Ohio soils. Weed Sci. 39: 186–194.

Cardina J, Sparrow DH and McCoy EL (1996) Spatial relationships between seedbank and seedling populations of common lambsquarters (*Chenopodium album*) and annual grasses. Weed Sci. 44: 298–308.

Cardina J, Johnson GA and Sparrow DH (1997) The nature and consequence of weed spatial distribution. Weed Sci. 45: 364–373.

Cardina J, Herms CP and Doohan DJ (2002) Crop rotation and tillage system effects on weed seedbanks. Weed Sci. 50: 448–460.

Cavers PB (1995) Seed banks: Memory in soil. Can. J. Soil Sci. 75: 11–13.

Clements DR, Benott DL, Murphy SD and Swanton CJ (1996) Tillage effects on weed seed return and seedbank composition. Weed Sci. 44: 314–322.

Colbach N, Forcella F and Johnson GA (2000) Spatial and temporal stability of weed populations over five years. Weed Sci. 48: 366–377.

Cromer HE, Murphy SD and Swanton CJ (1999) Influence of tillage and crop residue on postdispersal predation of weed seeds. Weed Sci. 47: 184–194.

Crow WD, Steckel LE, Hayes RM and Mueller TC (2015) Evaluation of POST-harvest herbicide applications for seed prevention of glyphosate-resistant Palmer amaranth (*Amaranthus palmeri*). Weed Technol. 29: 405–411.

Davis AS (2006) When does it make sense to target the weed seed bank? Weed Sci. 54: 558–565.

Davis AS and Liebman M (2003) Cropping system effects on giant foxtail (*Setaria faberi*) demography: I. Green manure and tillage timing. Weed Sci. 51: 919–929.

Davis VM, Gibson KD, Bauman TT, Weller SC and Johnson WG (2009) Influence of weed management practices and crop rotation on glyphosate-resistant horseweed (*Conyza canadensis*) population dynamics and crop yield-rears III and IV. Weed Sci. 57: 417–426.

DeVore JD, Norsworthy JK and Brye KR (2013) Influence of deep tillage, a rye cover crop, and various soybean production systems on Palmer amaranth emergence in soybean. Weed Technol. 27: 263–270.

Forcella F, Eradat-Oskoui K and Wagner SW (1993) Application of weed seedbank ecology to low-input crop management. Ecol. Appl. 3: 74–83.

Gallagher RS and Cardina J (1998) Phytochrome-mediated *Amaranthus* germination II: development of very low fluence sensitivity. Weed Sci. 46: 53–58.

Gallant ER (2006) How can we target the weed seedbank? Weed Sci. 54: 588–596.

Grimmer OP and Masiunas JB (2004) Evaluation of winter-killed cover crops preceding snap pea. Hort. Technol. 14: 349–355.

Gunsolus JL (1990) Mechanical and cultural weed control in corn and soybeans. Am. J. Alter. Agrc. 5: 114–119.

Jana S and Thai KM (1987) Patterns of changes of dormant genotypes in *Avena fatua* populations under different agricultural conditions. Can. J. Bot. 65: 1741–1745.

Jha P, Norsworthy JK, Bridges Jr W and Riley MB (2008) Influence of glyphosate timing and row width on Palmer amaranth (*Amaranthus palmeri*) and pusley (*Richardia* spp.) demographics in glyphosate-resistant soybean. Weed Sci. 56: 408–415.

Jha P and Norsworthy JK (2012) Influence of late-season herbicide applications on control, fecundity, and progeny fitness of glyphosate-resistant Palmer Amaranth (*Amaranthus palmeri*) biotypes from Arkansas. Weed Technol. 26: 807–812.

Jha P, Norsworthy JK, Riley MB and Bridges Jr W (2010) Annual changes in temperature and light requirements for germination of Palmer amaranth (*Amaranthus palmeri*) seeds retrieved from soil. Weed Sci. 58: 426–432.

Johnson III WC and Mullinix Jr BG (1995) Weed management in peanut using stale seedbed techniques. Weed Sci. 43: 293–297.

Johnson III WC and Mullinix Jr BG (1998) Stale seedbed weed control in cucumber. Weed Sci. 46: 698–702.

Jasieniuk M, Bru^lé-Babel AL and Morrison IN (1996) The evolution and genetics of herbicide resistance in weeds. Weed Sci. 44: 176–193.

Jones RE and Medd RW (2000) Economic thresholds and the case for long-term approaches to population management of weeds. Weed Technol. 14: 337–350.

Jordan N, Mortensen DA, Prenzlow DM and Cox KC (1995) Simulation analysis of crop rotation effects on weed seedbanks. Am. J. Bot. 82: 390–398.

Kegode GO, Forcella F and Clay S (1999) Influence of crop rotation, tillage, and management inputs on weed seed production. Weed Sci. 47: 175–183.

Knezevic SZ, Evans SP, Blankenship EE, Van Acker RC and Lindquist JL (2002) Critical period for weed control: the concept and data analysis. Weed Sci. 50: 773–786.

Kumar V and Jha P (2015a) Influence of herbicides applied postharvest in wheat stubble on control, fecundity, and progeny fitness of *Kochia scoparia* in US Great Plains. Crop Prot. 71: 144–149.

Kumar V and Jha P (2015b) Influence of glyphosate timing on *Kochia scoparia* demographics in glyphosate-resistant sugar beet. Crop Prot. 76: 39–45.

Leon RG and Owen MDK (2006) Tillage systems and seed dormancy effects on common waterhemp (*Amaranthus tuberculatus*) seedling emergence. Weed Sci. 54: 1037–1044.

Liebman M and Davis AS (2000) Integration of soil, crop and weed management in low-external-input farming systems. Weed Res. 40: 27–48.

Malik MS, Norsworthy JK, Culpepper AS, Riley MB and Bridges Jr W (2008) Use of wild radish (*Raphanus raphanistrum*) and rye cover crops for weed suppression in sweet corn. Weed Sci. 56: 588–595.

Mulugeta D and Stoltenberg DE (1997) Weed and seedbank management with integrated methods as influenced by tillage. Weed Sci. 45: 706–715.

Neve P, Vila-Aiub M and Roux F (2009) Evolutionary-thinking in agricultural weed management. New Phytol. 184: 783–793.

Neve P, Norsworthy JK, Smith KL and Zelaya IA (2011) Modelling evolution and management of glyphosate resistance in *Amaranthus palmeri*. Weed Res. 51: 99–112.

Norris RF (1999) Ecological implications of using thresholds for weed management. pp. 31–58. *In*: Buhler DD (ed.). Expanding the Context of Weed Management. Haworth Press, New York.

Norsworthy JK, McClelland M, Griffith G, Bangarwa SK and Still J (2011) Evaluation of cereal and brassicaceae cover crops in conservation-tillage, enhanced glyphosate-resistant cotton. Weed Technol. 25: 6–13.

Norsworthy JK (2003) Allelopathic potential of wild radish (*Raphanus raphanistrum*). Weed Technol. 17: 307–313.

Norsworthy JK, Malik MS, Jha P and Riley MB (2007) Suppression of *Digitaria sanguinalis* and *Amaranthus palmeri* using autumn-sown glucosinolate-producing cover crops in organically grown bell pepper. Weed Res. 47: 425–432.

Norsworthy JK, Ward SM, Shaw DR, Llewellyn RS, Nichols RL, Webster TM, Bradley KW, Frisvold G, Powles SB, Burgos NR, Witt WW and Barrett M (2012) Reducing the risks of herbicide resistance: best management practices and recommendations. Weed Sci. 60: 31–62.

Nurse RE, Hamill AS, Kells JJ and Sikkema PH (2008) Annual weed control may be improved when AMS is added to below-label glyphosate doses in glyphosate-tolerant maize (*Zea mays* L.). Crop Prot. 27: 452–458.

O'Donovan JT, Newman JC, Harker KN and Clayton GW (2004) Crop seeding rate influences the performance of variable herbicide rates in a canola–barley–canola rotation. Weed Technol. 18: 733–741.

O'Donovan JT, Blackshaw RE, Harker KN, Clayton GW, Moyer JR, Dosdall LM, Maurice DC and Turkington TK (2007) Integrated approaches to managing weeds in spring-sown crops in western Canada. Crop Prot. 26: 390–398.

Pareja MR, Staniforth DW and Pareja GP (1985) Distribution of weed seed among soil structural units. Weed Sci. 33: 182–189.

Price AJ, Balkcom KS, Culpepper SA, Kelton JA, Nichols RL and Schomberg H (2011) Glyphosate-resistant Palmer amaranth: a threat to conservation tillage. J. Soil Water Conserv. 66: 265–275.

Reddy KN and Koger CH (2004) Live and killed hairy vetch cover crop effects on weeds and yield in glyphosate-resistant corn. Weed Technol. 18: 835–840.

Riar DS, Ball DA, Yenish JP and Burke IC (2011) Light-activated, sensor-controlled sprayer provides effective postemergence control of broadleaf weeds in fallow. Weed Technol. 25: 447–453.

Roberts HA and Feast PM (1972) Fate of seeds of some annual weeds in different depths of cultivated and undisturbed soil. Weed Res. 12: 316–324.

Roggenkamp GJ, Mason SC and Martin AR (2000) Velvet leaf (*Abutilon theophrasti*) and green foxtail (*Setaria viridis*) response to corn (*Zea mays*) hybrid 1. Weed Technol. 14: 304–311.

Simard M, Rouane S and Leroux GD (2011) Herbicide rate, glyphosate/glufosinate sequence and corn/soybean rotation effects on weed seed banks. Weed Sci. 59: 398–403.

Shaw DR (1996) Development of stale seedbed weed control programs for southern row crops. Weed Sci. 44: 413–416.

Schillinger WF and Young FL (2000) Soil water use and growth of Russian thistle after wheat harvest. Agron. J. 92: 167–172.

Schreiber MM (1992) Influence of tillage, crop rotation, and weed management on giant foxtail (*Setaria faberi*) population dynamics and corn yield. Weed Sci. 40: 645–653.

Steckel LE, Sprague CL, Stoller EW, Wax LM and Simmons FW (2007) Tillage, cropping system, and soil depth effects on common waterhemp (*Amaranthus rudis*) seed-bank persistence. Weed Sci. 55: 235–239.

Shuma JM, Quick WA, Raju MVS and Hsiao AI (1995) Germination of seeds from plants of *Avenafatua* L. treated with glyphosate. Weed Res. 35: 249–255.

Swanton CJ and Booth BD (2004) Management of weed seedbanks in the context of populations and communities. Weed Technol. 18: 1496–1502.

Schweizer EE and Zimdahl RL (1984) Weed seed decline in irrigated soil after six years of continuous corn (*Zea mays*) and herbicides. Weed Sci. 32: 76–83.

Taylor SE and Oliver LR (1997) Sicklepod (*Senna obtusifolia*) seed production and viability as influenced by late-season post-emergence herbicide applications. Weed Sci. 45: 497–501.

Taylor KL and Hartzler RG (2000) Effect of seed bank augmentation on herbicide efficacy. Weed Technol. 14: 261–267.

Teasdale JR, Beste CE and Potts WE (1991) Response of weeds to tillage and cover crop residue. Weed Sci. 39: 195–199.

Teasdale JR and Daughtry CS (1993) Weed suppression by live and desiccated hairy vetch (*Vicia villosa*). Weed Sci. 41: 207–212.

Teasdale JR and Mohler CL (1993) Light transmittance, soil temperature, and soil moisture under residue of hairy vetch and rye. Agron. J. 85: 673–680.

Teasdale JR, Brandsaeter LO, Calegari A and Neto FS (2007) 4 cover crops and weed management. pp. 49–64. *In*: Upadhyaya MK and Blackshaw RE (eds.). Non-chemical Weed Management: Principles, Concepts, and Technology. CAB International, Oxfordshire, UK.

Vencill WK, Nichols RL, Webster TM, Soteres JK, Mallory-Smith C, Burgos NR, Johnson WG and McClelland MR (2012) Herbicide resistance: toward an understanding of resistance development and the impact of herbicide-resistant crops. Weed Sci. 60: 2–30.

Wiggins MS, McClure A, Hayes RM and Steckel LE (2015) Integrating cover crops and POST herbicides for glyphosate-resistant Palmer amaranth (*Amaranthus palmeri*) control in corn. Weed Technol. 29: 412–418.

Werth JA, Preston C, Roberts GN and Taylor IN (2008) Weed management impacts on the population dynamics of barnyardgrass (*Echinochloa crus-galli*) in glyphosate-resistant cotton in Australia. Weed Technol. 22: 190–194.

Walsh MJ and Powles SB (2007) Management strategies for herbicide-resistant weed populations in Australian dryland crop production systems. Weed Technol. 21: 332–338.

Walsh M and Newman P (2007) Burning narrow windrows for weed seed destruction. Field Crops Res. 104: 24–30.

Walsh MJ, Harrington RB and Powles SB (2012) Harrington seed destructor: A new nonchemical weed control tool for global grain crops. Crop Sci. 52: 1343–1347.

Walsh M, Newman P and Powles S (2013) Targeting weed seeds in-crop: A new weed control paradigm for global agriculture. Weed Technol. 27: 431–436.

Walsh MJ and Powles SB (2014) High seed retention at maturity of annual weeds infesting crop fields highlights the potential for harvest weed seed control. Weed Technol. 28: 486–494.

White SS, Renner KA, Menalled FD and Landia DA (2007) Feeding preferences of weed seed predators and effect on weed emergence. Weed Sci. 55: 606–612.

Wilson RG and Sbatella GM (2011) Late-season weed control in glyphosate-resistant sugar beet. Weed Technol. 25: 350–355.

Wilson RG, Young BG, Matthews JL, Weller SC, Johnson WG, Jordan DL, Owen MDK, Dixon PM and Shaw DR (2011) Benchmark study on glyphosate-resistant cropping systems in the United States. Part 4: Weed management practices and effects on weed populations and soil seedbanks. Pest Manage. Sci. 67: 771–780.

Yenish JP, Doll JD and Buhler DD (1992). Effects of tillage on vertical distribution and viability of weed seed in soil. Weed Sci. 40: 429–433.

Young FL (1986) Russian thistle (*Salsola iberica*) growth and development in wheat (*Triticum aestivum*). Weed Sci. 34: 901–905.

Chapter 3

Effect of Environmental Factors on Weed Germination and Management

Neha Rana,[1,] Aruna V. Varanasi[2] and Brent A. Sellers[3]*

Introduction

Weeds are major constraints for successful crop production. One of the most important characteristics of weeds is their ability to adapt and survive under various habitats. Weeds show a greater range of responses to environmental conditions due to their broader gene pool (Treharne 1989). In addition to high genetic diversity, weeds possess several other traits such as early seedling emergence, rapid growth rate, and prolific seed production that make them strong competitors to crops. Due to these characteristics, weeds can survive and outgrow other species in an ecological niche. Weeds with rapid seed dispersal and prolonged emergence periods can quickly spread into new territories and alter the seedbank composition and weed population dynamics. If left unchecked, weeds emerging in high densities with or prior to crop emergence strongly compete with crops resulting in significant yield losses.

As with all plants, seed germination and emergence play an important role in weed establishment and competitiveness. Seed germination and

[1] Research Scientist, Monsanto Company, 700 Chesterfield Parkway W, Chesterfield, MO 63017.
[2] Postdoctoral Research Associate, 3723 Throckmorton Ctr., Kansas State University, Manhattan, KS 66506.
[3] Associate Professor, Range Cattle Research and Education Center, University of Florida, 3401 Experiment Station, Ona, FL 33865.
* Corresponding author: Neha.rana@monsanto.com

emergence is a crucial phase in plant development that determines the population dynamics in a particular habitat (Forcella et al. 2000). Uncontrolled weeds continue growth and reproduction and deposit most of their seeds back to the soil and increase their seed numbers in the soil (weed seedbank) rapidly from year to year. Regeneration and persistence of weeds in a field is mainly influenced by the replenishment of weed seedbank and seedling establishment. Weeds produce viable seeds that can germinate and establish relatively quickly compared to crops even under environmental and soil conditions that are unfavorable for crop growth. Furthermore, most weeds can induce dormancy under unfavorable growth conditions and initiate germination only when environmental conditions are adequate, thus preventing seed decay and loss. Weeds are most sensitive at the seedling stage and weed management strategies focused to control weeds at this stage ensure for greater weed suppression throughout the crop growth. Hence, timely weed control during the early stages of crop growth is crucial to sustaining crop productivity.

Herbicides are the major and most effective tools used for weed control. However, due to environmental concerns and increasing reports of development of herbicide resistance in several weeds (Heap 2016), there is growing pressure to design alternative weed management strategies that are more efficient and sustainable. Optimizing cultural methods such as tillage, seed bed preparation, and mulching to reduce weed emergence and weed seedbank deposition is one such alternative (Rana et al. 2012, 2015; Grundy et al. 2003). To develop such alternative strategies, it is essential to understand the biological, ecological, and environmental factors that affect the behavior and persistence of weed seeds under field conditions. This knowledge will also assist in developing models that can predict the weed invasion potential and weed seed bank dynamics (Kriticos et al. 2003). Models on the composition of weed seedbank will help estimate persistence of weed populations and the resulting crop yield losses. This chapter provides an overview of current knowledge on the seed behavior and factors affecting seed germination and emergence in weeds. Critical information on the favorable environments needed for weed germination and establishment is also reviewed. Such knowledge will be useful for developing improved weed management strategies.

Factors affecting Weed Seed Germination and Emergence

Seed dormancy and germination are complex processes regulated by both endogenous and environmental factors before and after seed development (Koorneef et al. 2002; Finch-Savage and Leubner-Metzger 2006). Weed seed establishment is influenced by a number of factors that may be present inherently in the seed (physiological factors) or mediated by various environmental factors such as light, temperature, moisture, oxygen, and burial depth (Chauhan et al. 2006; Nandula et al. 2006; Li et al. 2012; Li et al. 2015). The interaction among these factors further complicates the germination process and seedling performance. Therefore, a better understanding of the species-specific seed germination response to these factors is essential to develop appropriate and timely weed control measures tailored to the composition of weed species in an ecological niche.

Physiological Factors

Seed biology

The biological characteristics of seeds that influence seed germination include seed size, seed shape, seed mass, and seed coat structure. These characteristics vary widely among weed species and, in some cases, among various populations of the same species (Van Molken et al. 2005; Tanveer et al. 2013). Seed polymorphism (distinctly different types of seeds produced by a species) is an important survival mechanism in many species used to cope with differences in microhabitats related to both edaphic and environmental conditions (Imbert 2002). Information about the effects of these factors is crucial to develop weed control strategies that prevent the spread and establishment of weed seeds with a wide variety of biological traits.

Variation in seed size may result in variability in seed germination characteristics, emergence traits, and persistence in weed seedbank under different environments (Bekker et al. 1998). Heterogeneous seed germination due to seed size variation is an adaptive trait found in several species which allows the plant populations to expand their geographic distribution and flourish in various types of habitats. The influence of seed size on germination and seedling performance has been documented in several plant species (Willenborg et al. 2005; Li and Feng 2009; Saatkamp et al. 2009; Tanveer et al. 2013). Besides influencing seed germination, seed size also impacts dispersal mechanisms and spread into new areas. Larger and heavier seeds may result in reduced dispersibility and slower germination compared to smaller and lighter seeds (Bakker et al. 1996; Imbert and Ronce 2001).

Small-seeded weeds like pigweeds (Amaranthus spp.), common purslane (*Portulaca oleraceae*), common lambsquarters (*Chennopodium album*), and many grass weeds exhibit prolific seed production within a single season, but have limited carbohydrate reserves and need to remain near the soil surface (shallow depths) to be able to germinate, develop and reproduce (Kon et al. 2007). This may also be a disadvantage as the seeds may be subjected to decay by microorganisms or lost to natural predation or extreme temperatures. Furthermore, seedlings from small seeds may produce less dense tissue and become more sensitive to physical damage (Grundy et al. 2003). However, when conditions are adequate, small seeds germinate faster and grow rapidly to exploit their opportunities to establish and compete with other species in the vicinity (Leishman et al. 2000). Small seeds buried deep into the soil tend to evade germination and induce dormancy as a survival mechanism until conditions are suitable to promote germination. Although some seeds loose viability because of decay, high seed production enables these weeds to emerge in multiple flushes throughout the growing season which further complicates weed control (Kon et al. 2007). Agronomic practices like tillage may bring the buried seeds closer to the soil surface and indirectly promote germination (Harrison et al. 2007). This is an important implication for weed management, as tillage can be used to either encourage germination so as to suppress weed growth completely before crop planting or to bury them deeper from where they cannot germinate and emerge in high densities (Froud-Williams et al. 1983).

In contrast to small seeds, large-seeded species such as common cocklebur (*Xanthium strumarium*), morningglory (*Ipomoea* spp.), giant ragweed (*Ambrosia trifida*), and velvetleaf (*Abutilon theophrasti*) have high amounts of carbohydrate reserves and can germinate from greater burial depths (Baskin and Baskin 1998; Grundy et al. 2003). Deeper burial helps these species to prevent seed decay and loss, thereby allowing them to persist in soil for longer periods (Bararpour and Oliver 1998). Although they germinate slowly, and seed production is less prolific, higher germination percentage and seedling establishment ensures their survival and growth in various habitats. Seed size was reported to have adverse effects on seed germination and emergence traits of field bindweed (*Convolvulus arvensis*) (Tanveer et al. 2013). Larger seeds tend to have higher germination percentage and improved seedling establishment compared to small seeds, irrespective of moisture/salt stress or deeper burial conditions. Harrison et al. (2007) observed that variation in seed size significantly interacted with burial depth in giant ragweed. Seed size had no effects on variability in germination when placed near the soil surface or buried at 20 cm depth. However, the probability of emergence varied between small seeds (9% germination) and large seeds (30% germination) at a burial depth of 10 cm. At this depth, large seeds will have greater emergence percentage which ensure seedling establishment and higher plant vigor. On the other hand, even though smaller seeds produce fewer seedlings at this soil depth, they will have lower viability losses and thus, are important for extending seed persistence in the soil. These findings suggest that the polymorphism in seed size may facilitate adaptation across different habitats in some weed species and is an important factor to consider when devising weed control strategies.

Seed mass and shape have also been suggested to play an important role in influencing seed germination and seedling behavior. The interactions of these morphological characteristics with burial depth were found to be complex and species-specific (Thompson et al. 1993; Grundy et al. 2003). Seed mass is an important measure of seed germination as it determines the resources available to the seedling during emergence (Westonby 1998). Seed mass also affects the weed's ability to emerge from deeper layers of soil and continue to grow under adverse conditions. While seed mass is a measure of seed's inherent resources, seed shape (seed surface area that is in contact with the external environments) impacts the resource capturing ability (e.g., moisture absorption, light perception). Seed shape may also influence the ability of seeds to penetrate into the soil and persist for longer periods. For example, flattened seeds (seeds of many grass weeds) have a lower tendency to penetrate the soil and bury into greater depths compared to rounded seeds and tend to remain on the soil surface. Although they have greater contact with the underlying soil, these seeds are typically short-lived and result in transient seedbanks as opposed to seeds with other shapes (Thompson et al. 1993; Moles et al. 2000; Traba et al. 2004).

Seed coat (outer layer of a mature seed) is another important determinant of seed germination and longevity. It is an essential component of seed structure that mediates the relationship between seed and its environment. Seed coats determine the thickness of seeds and weed species vary greatly in the structure and composition of seed coats. Seed coat influences essential functions such as embryo and endosperm preservation, protection of internal structures of the seed from external damages, and regulation of dormancy (Souza and Marcos-Filho 2001; Moise et al. 2005). In addition, seed coats contain phenolic compounds that contribute to resistance to biotic and abiotic stress and may also a play a major role in directing the nutrient supply to embryo during germination (Weber et al. 2005; Davis et al. 2008). Seed coat thickness determines the ability of seeds to remain dormant until embryo maturation and certain environmental fluctuations such as alternating hot and cold temperatures, enable germination (Bewley 1997). Seed coat thickness has been suggested as one of the important determinants of seed longevity in the soil for arable weeds (Gardarin et al. 2010). In general, species with thin seed coats [kochia (*Kochia scoparia*), and most annual grass weeds] germinate faster than those with thicker seed coats [redroot pigweed (*Amaranthus retroflexus*), Palmer amaranth (*Amaranthus palmeri*), wild mustard (*Sinapis arvensis*), field pepperweed (*Lepedium campestre*)] but have less persistence in the soil seedbanks as they are mostly non-dormant. Weed species with hard seed coats contribute largely to the accumulation of weed seedbanks in cultivated fields due to their ability to resist mechanical and environmental stress for longer periods. Moreover, weeds with hard seed coats may result in prolonged emergence periods resulting in multiple flushes throughout the crop growing season, potentially causing interference in harvest operations and greater yield losses. Gardarin et al. (2010) studied the relationship between seed coat thickness and seed mortality in 13 weed species and reported that seed coat thickness varied significantly (17 to 231 µm) among these weed species. Seed decay and loss decreased with increase in seed coat thickness, thus confirming it to be an important trait for measuring seed mortality rates in the soil. Weed species such as common poppy (*Papaver rhoeas*), sun spurge (*Euphorbia helioscopia*), and some weeds of Orobanchaceae family have an alveolar (smooth) seed coat that enables them to bury deeper and faster, and tend to create persistent seedbanks (Roberts and Feast 1973; Lopez-Granados and Garcia-Torres 1999).

Seed dormancy

Dormancy is a process when viable seeds enter into a period of a resting stage, lasting for few days, weeks or even years, to prevent germination, thereby preventing seed loss under unfavorable growth conditions (Bewley 1997; Finch-Savage and Leubner-Metzger 2006). Seed dormancy is an important survival mechanism in several weed species. This mechanism is especially common in summer annual weeds such as Palmer amaranth, common lambsquarters, velvetleaf, and most grass weeds to enable germination only when seed and environmental conditions are adequate (Gulden and Shirtliffe 2009). Although seed dormancy is considered as an undesirable trait in agricultural crops, dormancy regulation is important in weeds to optimize germination timing and facilitate distribution of weed populations over time (Bewley 1997). Dormancy is an important adaptive trait that allows weed populations to delay germination and persist in the soil for long periods until favorable conditions return. For instance, winter annual weeds produce seeds in the spring but plants may not survive the heat in summer and therefore, induce dormancy to prevent germination in summer conditions. However, dormancy is broken and seeds are able to germinate in the fall when conditions are favorable.

Extensive work has been published on dormancy regulation and seed germination in crops and weed species (Baskin and Baskin 1998, 2004; Bewley 1997; Benech-Arnold et al. 2000; Fenner and Thompson 2005; Gardarin and Colbach 2015). Seed dormancy is determined by a number of factors including environmental (light, temperature, burial depth), structural (seed coat thickness), physiological (plant hormones and embryo maturation), and agronomic practices (tillage, irrigation, cropping systems) (Donohue et al. 2005; Huang et al. 2010).

Dormancy in weeds can be induced by several mechanisms. Seeds that are immature at the time of shedding typically require an after-ripening period to fully develop and initiate germination (Holdsworth et al. 2008). This type of dormancy is called innate or primary dormancy and is commonly found is some grass weeds like foxtail (*Setaria* spp.) and wild oat (*Avena fatua*). Innate dormancy is caused by immature embryos, hard seed coat, or germination inhibitors. Hard seed coat dormancy typically depends on the balance between the force exerted by the radicle and the strength of the seed coat (Graeber et al. 2012). This dormancy can be released by processes that mechanically remove the seed coat to promote germination (scarification) or by exposure to fluctuating temperatures that soften the seed coat (stratification). In some weeds, dormancy can be induced temporarily due to extreme environmental conditions such as hot or cold temperatures or high carbon dioxide levels at the root zone. This type of dormancy known as induced dormancy prevents germination during unfavorable growth conditions and is quickly broken when conditions return to normal. For example, high temperatures induce dormancy in summer annual weeds such as yellow foxtail (*Setaria pumila*) and redroot pigweed, to avoid germination in the fall and growth in winter conditions. Seeds remain dormant until spring and germinate as soon as temperatures are optimum. Enforced dormancy is another type of dormancy that occurs due to lack of specific environmental condition such as cooler temperatures, optimum moisture and salinity, or lack of light and oxygen when buried deep into the soil. Germination is induced when the environmental condition is fulfilled, thereby removing the limitation that induced dormancy. For example, seeds of common chickweed (*Stellaria media*) will not germinate even when all other conditions are favorable as they require cold temperatures to initiate germination. While innate dormancy, once terminated, usually does not recur, induced or enforced dormancy (also known as secondary dormancy) may repeat in the same set of viable seeds whenever conditions are not optimum, thus enabling higher germination potential throughout the growing season (Footitt et al. 2011).

Seed dormancy and germination are complex heritable traits determined by multiple genetic factors. Several plant hormones such as abscisic acid (ABA), gibberellins (GAs), ethylene, and brassinosteroids (BR) play a major role in dormancy regulation. ABA and GA are the key hormones that have antagonistic roles in regulating the equilibrium between dormancy and germination (Finkelstein et al. 2008; Rodriguez-Gacio et al. 2009). ABA is a positive regulator of dormancy and its signaling is essential to induce and maintain dormancy in the seeds (Kucera et al. 2005; Finch-Savage and Leubner-Metzger 2006). In contrast, GA has been shown to promote growth of the embryo and initiate germination (Holdsworth et al. 2008; Matilla and Matilla-Vázquez 2008; Linkies and Leubner-Metzger 2012). The role of GA during dormancy release and germination is well documented (Sun and Gubler 2004; Yamauchi et al. 2004, 2007; Liu et al. 2014). In response to environmental signals such as optimum light and temperature, imbibed seeds tend to decrease ABA content and stimulate GA biosynthesis to terminate dormancy and induce germination. The dynamic nature of ABA:GA ratios largely determine dormancy and germination processes in plants. High ABA:GA ratios are required to maintain dormancy, whereas dormancy release and germination initiation involves increased GA biosynthesis with simultaneous degradation of ABA (Ali-Rachedi et al. 2004; Cadman et al. 2006). It is suggested that GA may not directly involve in dormancy release but is only needed in sufficient concentrations to induce germination as soon as ABA is degraded (Le Page-Degivry et al. 1997). This theory was further supported genetically by Fennimore and Foley (1998) who demonstrated that addition of GA to the growth medium only induced germination in dormant seeds of wild oat and did not directly affect dormancy termination. Other hormones, such as ethylene and BR also counteract ABA effects and promote germination. Although produced in trace amounts, ethylene is involved in several plant growth processes ranging from germination to senescence. Ethylene response is more often associated with seeds that are light sensitive. Ethylene is known to interact with GA and other hormones to interfere in ABA signaling and promote germination (Kepczynski et al. 1997; Matilla 2000; Kucera et al. 2005). Interactions between ABA and ethylene signaling during germination can significantly impact emergence and early seedling growth (Ghassemian et al. 2000; Chiwocha et al. 2005). BR also acts synergistically with GA to promote cell elongation and germination in some species (Steber and McCourt 2001).

Environmental Factors

The environmental or abiotic factors such as light, temperature, water stress, depth of burial, and oxygen also play a critical role in weed seed germination. Each species has its own precise requirement to initiate the germination process; deviation from these requirements result in failure of seed emergence. So it is important to understand the environmental conditions of the weed seed to better manage them. Light quality and quantity may trigger or inhibit germination depending on the weed seed requirement. Germination of buried seeds is held in check by light and mositure. Similarly, different levels of oxygen concentration in soil can result in different levels of germination. These factors are discussed in more detail below.

Light

Light is one of the most essential requirements for seeds to germinate (Wesson and Wareing 1967). Mostly, the seed that require light for germination exhibit physical or seed coat-imposed dormancy. Photoperiod can impact seed dormancy in several species. For example, common lambsquarters seed that mature under long days are smaller with thicker seed coat, while those develop in short days are non-dormant (Karssen 1970). Phytochromes, the family of chromoproteins, also regulate seed germination in response to light (Thompson and Grime 1983). Phytochrome exists in two photo-interconvertible forms: Pr (red form) and Pfr (far-red form). Pr absorbs red light (peak at 660 nm) and is immediately converted to biologically active Pfr, resulting in seed germination (Hennig et al. 2002; Shinomura et al. 1994). Pfr absorbs far-red light (peak at 730 nm) and is quickly converted back to biologically-inactive form (Pr). Dormancy is terminated if the light source remains as Pfr, but is retained if Pr form is continued. Thus, the phytochrome acts as a biological light switch to initiate seed germination. In weeds that are deeply buried, some weeds display extreme sensitivity to light known as very low fluence (VLF) response (Batlla and Benech-Arnold 2014). In such cases, even a brief exposure to light is enough to induce germination (Scopel et al. 1991; Botto et al. 1996). Common chickweed and knotgrass (*Polygonum aviculare*) have shown to exhibit VLF response (Taylorson 1972; Batlla and Benech-Arnold 2005). Palmer amaranth seed buried in spring, resumed germination upon a brief exposure to red light in autumn (Jha et al. 2010). However, in some weed species like curly dock (*Rumex crispus*) germination may occur close to soil surface or on soil surface in the absence of any vegetation because of high red to far-red ratio (R:FR) of sunlight rather than brief exposure to light (Deregibus et al. 1994; Insausti et al. 1995). On contrary, in Italian ryegrass (*Lolium multiflorum*) a small decrease in R:FR ratio is enough to reduce germination (Deregibus et al. 1994; Batlla et al. 2000).

The requirement of light for seed germination also depends on temperature. Light interacts with various temperature fluctuations to terminate dormancy, and thus enables germination. In pigweed Pfr light cannot break dormancy at high temperature (40 C), but a temperature drop to 26 C for about 1 h after exposure to Pfr helps terminate dormancy (Bewley and Black 1994). On the other hand, light had minimal impact on dormancy of pitted morningglory (*Ipomea lacunosa*) and tall morningglory (*Ipomea purpurea*) (Norsworthy and Oliveira 2007; Jha et al. 2015). Germination in both these species is independent of light and requires temperature alternations to break physical dormancy and induce germination. From a management standpoint, these factors become harder to control because seeds are able to germinate on either bare or canopy-covered soil.

Temperature

Temperature affects both the ability and rate of weed seed germination (Baskin and Baskin 1977; Rana et al. 2012). Temperature plays a major role in removing primary/secondary dormancy but sometimes can also induce secondary dormancy to regulate germination under field conditions. Recently, thermal adaptations of horseweed (*Conyza Canadensis*) to different environments have been investigated (Tozzi et al. 2013). Different base temperature regimes are needed for 50% germination of horseweed under different geographies. The role of high temperature in devitalizing weed seed has been reported by Vidotto et al. (2013). Barnyard grass (*Echinochloa crusgalli*), green foxtail (*Setaria viridis*), black night shade (*Solanum nigrum*), common purslane, and redroot pigweed are sensitive to high temperature exposure indicating that seed size and weight was directly proportional to heat tolerance.

Common ragweed (*Ambrosia artemisiifolia*), common lambsquarters, Palmer amaranth, and common waterhemp (*Amaranthus rudis*) are some of the most troublesome weeds in the US Midwestern row-crops (WSSA Weeds Surveys 2016). Temperature has a prevalent effect on germination of these weed species. Emergence of Amaranthus spp. is susceptible to low temperature stress and the emergence declines above 35/30 C (day/night temperatures) (Guo and Al-Khatib 2003). Similarly, germination was higher in forage kochia (*Kochia prostrata*) at 20 C than at 10 and 30 C (Romo and Haferkamp 1987). Common ragweed and common lambsquarters germinate in early spring while redroot pigweed germinates from late spring to early summer (Baskin and Baskin 1977). Redroot pigweed prefers > 25 C, whereas, common lambsquarters and common ragweed require > 10 C and > 15 C, respectively for germination (Bewley and Black 1994; Ghorbani et al. 1999). In kochia dormancy is terminated by chilling and seeds start to germinate when temperature begins to rise in spring (Bewley and Black 1994). This has a dual advantage; firstly, emergence before inclement weather is prevented and secondly, emergence occurs early enough to establish at a favorable time when there is no shade from crops.

Interestingly, seeds of tall morning glory are more dormant after 12 months of after-ripening compared to freshly matured seed (Jha et al. 2015). Seeds go through a 30 C constant or 22.5/37.5 C temperature alternations in winter for 3 months after maturation, then the after-ripened seed stay on soil surface for 6 months to germinate at 10–40 C. Exposure to hot summer temperatures induce secondary dormancy in these seed, and only a small fraction of non-dormant seed germinate in late summer or autumn at high temperature of 30–35 C. Therefore, management of tall morning glory is closely linked to dormancy termination and induction, which differs from year to year because of its dependency on environmental conditions.

Water stress

Soil moisture affects both the timing and rate of weed seed emergence (Boyd and Van Acker 2003, 2004). Seed germination can be affected under dry or high-moisture conditions. In dry conditions, seed germination is prevented, whereas, under high-moisture conditions, hypoxia or the inability to remove toxic compounds may limit seed germination (Roberts and Potter 1980; Holm 1972). Soil moisture availability is altered by many variables including tillage, cover crops, litter, and soil texture (Scopel et al. 1994; Teasdale and Mohler 1993). The impact of water stress on weed seed emergence is highly variable because moisture conditions within a field may vary considerably. Barnyard grass is an unique example where germination declines at moisture stress –0.01 MPa and increases at –0.5 MPa and is sensitive to oxygen levels (Boyd and Van Acker 2004). In contrast, germination of green foxtail is significantly higher at –0.01 MPa than –0.5 MPa and relatively insensitive to oxygen concentration (Boyd and Van Acker 2004). However, germination of wild oat depends on seed exposure to light and water deficit (Hou and Simpson 1991, 1993).

Depth of burial

Soil depth is known to inhibit seed germination (Holm 1972). Ecological factors including reduced gas exchange, light penetration, temperature, and need of mechanical abrasion or degree of soil compaction play a vital role in limiting seed germination (Benvenuti and Macchia 1995; Wesson and Wareing 1967; Pareja and Staniforth 1985). The biological reason for failure of buried weed seed to germinate is not well understood. Of all 20 weed species tested by Benvenuti et al. (2001), johnsongrass (*Sorghum halepense*), velvetleaf, and cutleaf geranium (*Geranium dissectum*) germinated even from a depth of 10 cm. Seed emergence is slightly decreased at shallow burial of 2 cm, but with increasing burial depth the decrease in emergence was exponential; no weed species can germinate at depths greater than 12 cm (Benvenuti et al. 2001). In general weed seed perceive burial depth as an unfavorable condition for germination and respond by inducing depth-mediated dormancy rather than germination suicide (Wesson and Wareing 1969). This is an important survival strategy that allows seedbank accumulation and ultimately persistence of seedbank.

Light penetration is one of the most important factors that limits germination of buried seed (Woolley and Stoller 1978). Most of the weed seed emergence occurs from the uppermost layer of soil, which is a reservoir of large populations of buried seed (Benvenuti et al. 2001). These seed remain viable in soil and readily emerge when there is any soil disturbance by cultivation or tillage. This is possible due to the very short exposures to sunlight when the soil is mechanically disturbed (Scopel et al. 1994). Emergence of weeds such as wild mustard and prostrate knotweed (*Polygonum aviculare*) increase from 35 to 780 seedlings per square meter after tillage (Bewley and Black 1994). In a recent report, seedling recovery of large-seeded species from a complete burial at 2 cm soil depth was higher than small-seeded species under daily water regime (Mohler et al. 2016). This can influence weed management, because seedling recovery of large-seeded weed species can be minimized by withholding irrigation and maximizing burial depth.

Light penetrates the top few millimeters of the soil which induces germination of most light-sensitive seeds (Woolley and Stoller 1978). Common lambsquarters can germinate from a depth of < 2 mm in sand, but cannot germinate from 4 mm soil depth, while broadleaf dock (*Rumex obtusifolius*) can emerge from a depth of 4 mm, but cannot emerge from 6 mm soil depth (Batlla and Benech-Arnold 2014). The depth of light penetration depends on particle size, moisture content, and color of the soil. Longer wavelengths like far-red form penetrate deeper than shorter wavelengths (red form) (Baskin and Baskin 1998). The emergence of weed seed from different depths also depends on physical parameters of soil (Cussans et al. 1996). Depth-mediated germination inhibition has been reported to be directly proportional to clay content and inversely proportional to sand content in the soil (Benvenuti 2003). Weed seed tend to germinate from a wider range of depths in sandy soils and are less suited for long term accumulation, whereas soils with higher clay content are characterized by low weed seed emergence but tend to accumulate a persistent seedbank over time. Boyd and Van Acker (2003) studied how seeding depth and fluctuation in moisture level affect weed seed germination and emergence. Interestingly, surface seed of catch weed bedstraw (*Galium aparin*), wild oat, spring wheat (*Triticum aestivum*), and green foxtail had lower emergence compared to those at a depth of 1 to 4 cm when moisture levels were held constant or fluctuated.

Oxygen

Seeds require adequate aeration for germination. Oxygen requirements may depend on permeability of seed coat and the demands of the embryo. It is well known that limited oxygen supply can induce dormancy in some species, while complete absence of oxygen can result in the death of seeds. Soil moisture and depth of burial can cause the oxygen levels to vary in the soil profile. High levels of oxygen at the time when seeds are imbibing water prior to germination can also result in injury.

Oxygen concentration is inversely proportional to depth of seed burial (Topp et al. 2000). Lack of oxygen or hypoxia results in decreased seed germination with increasing soil depth (Holm 1972; Benvenuti and Macchia 1995). Soil air at a depth of six inches is similar in composition to above ground air (Russell and Appleyard 1915). Most seed tend to geminate at 15% oxygen and rate of germination increases with increase in oxygen concentrations (Gutterman et al. 1992). Wesson and Wareing (1969) were one of the few early pioneers who postulated presence of gas inhibitors with the onset of physiological processes of seed germination. Production of toxic volatile anaerobic metabolites such as acetaldehyde, acetone, and ethanol were identified in the seed when the oxygen levels surrounding the seed reached low levels (approx. 5%). The inhibitory effects of low oxygen concentration on seed germination of velvetleaf, morning glory, and wild mustard are alleviated by flushing the soil with air (Benvenuti and Macchia 1995; Holm 1972). It is important to note that the germination inhibition is not directly attributed to preexisting soil hypoxia, but rather to the hypoxic conditions created when consumption of oxygen increases due to increase in seed respiration rate (Benvenuti 2003). Therefore, poor soil gas diffusion due to low oxygen concentration or the inability to remove volatile toxic fermentation products surrounding the buried seed is responsible for inhibiting seed germination.

Conclusion

Both physiological and environmental factors govern germination and weed seed emergence. In addition to water and oxygen supply, optimum chemical environment inside the seed is essential to induce germination. Furthermore, physical environment including temperature, light, and soil moisture must be favorable. Even when all these conditions are satisfied, the seeds may still fail to germinate due to dormancy. Dormancy is an important adaptive mechanism to ensure seedling emergence at the best time and place. For a seed to release dormancy, it must go through certain environmental conditions or metabolic changes. Seed size, shape, and mass are vital characteristics that regulate seed germination. Smaller sized seeds with thin seed coats and round shape tend to germinate faster because they typically remain on the top soil, imbibe water faster, and germinate as soon as they perceive favorable environmental conditions. On the other hand, larger seeds with higher unit weight, thicker seed coat, and flattened seed shape can penetrate and germinate from deeper layers within the soil profile. Similarly, light quality, quantity and transmittance in soil is also important to induce or release dormancy and even to facilitate germination. Oxygen concentrations in the soil together with moisture stress can act as a signal preventing germination under dry conditions.

Knowledge of environment impact on seed germination is critical from a weed control perspective, because it helps to predict the timing and rate of seed germination for optimum PRE and POST herbicide application. Improved understanding of seed biology as well as physiological and environmental conditions regulating germination provide an opportunity for strategic management of weeds in various cropping systems. In the current era of weed resistance to herbicides, knowledge of seed biology and germination will assist immensely to develop weed management strategies related to tillage practices, crop rotation, and cover crops. It is crucial to understand how the physiological and environmental factors intertwine and affect each other during weed seed germination for efficient weed management.

LITERATURE CITED

Ali-Rachedi S, Bouinot D, Wagner MH, Bonnet M, Sotta B, Grappin P and Jullien M (2004) Changes in endogenous abscisic acid levels during dormancy release and maintenance of mature seeds: studies with the Cape Verde Islands ecotype, the dormant model of *Arabidopsis thaliana*. Planta 219: 479–488.

Bakker JP, Poschlod P, Strykstra RJ, Bekker RM and Thompson K (1996) Seed banks and seed dispersal: important topics in restoration ecology. Acta Botanica Neerlandica 45: 461–490.

Barapour MT and Oliver LR (1998) Effect of tillage and interference on common cocklebur (*Xanthium strumarium*) and sicklepod (*Senna obtusifolia*) population, seed production and seed bank. Weed Sci. 46: 424–431.

Baskin CC and Baskin JM (1998) Ecology of seed dormancy and germination in grasses. Population Biology of Grasses 28: 30–83.

Baskin JM and Baskin CC (1977) Role of temperature in the germination ecology of three summer annual weeds. Oecologia (Berl.) 30: 377–382.

Baskin JM and Baskin CC (2004) A classification system for seed dormancy. Seed Sci. Res. 14: 1–16.

Batlla D and Benech-Arnold RL (2005) Changes in the light sensitivity of buried *Polygonum aviculare* seeds in relation to cold-induced dormancy loss: development of a predictive model. New Phytol. 165: 445–452.

Batlla D and Benech-Arnold RLB (2014) Weed seed germination and the light environment: Implication for weed management. Weed Biol. Manag. 14: 77–87.

Batlla D, Kruk BC and Benech-Arnold RL (2000) Very early detection of canopy presence by seeds through perception of subtle modifications in red: far red signals. Funct. Ecol. 14: 195–202.

Bekker RM, Bakker JP, Grandin U, Kalamees R, Milberg P, Poschlod P, Thompson K and Willems JH (1998) Seed size, shape and vertical distribution in the soil: indicators of seed longevity. Functional Ecol. 12: 834–42.

Benech-Arnold RL, Sanchez RA, Forcella F, Kruk BC and Ghersa CM (2000) Environmental control of dormancy in weed seed banks in soil. Field Crops Res. 67: 105–122.

Benvenuti S (2003) Soil texture involvement in germination and emergence of buried weed seeds. Agron. J. 95: 191–198.

Benvenuti S and Macchia M (1995) Effect of hypoxia on buried weed seed germination. Weed Res. 35: 343–351.

Benvenuti S, Macchia M and Miele S (2001) Quantitative analysis of emergence of seedlings from buried weed seeds with increasing soil depth. Weed Sci. 49: 528–535.

Bewley JD (1997) Seed germination and dormancy. The Plant Cell 9: 1055–1066.

Bewley JD and Black M (1994) Seeds: Physiology of Development and Germination. 2nd ed. New York: Plenum Press. pp. 262–290.

Botto JF, Sanchez RA and Casal JJ (1998) Burial conditions affect light responses of *Datura ferox* seeds. Seed Sci. Res. 8: 423–429.

Botto JF, Sanchez RA, Whitelam GC and Casal JJ (1996) Phytochrome A mediates the promotion of seed germination by very low fluences of light and canopy shade light in Arabidopsis. Plant Physiol. 110: 439–444.

Boyd NS and Van Acker RC (2003) The effects of depth and fluctuating soil moisture on the emergence of eight annual and six perennial plant species. Weed Sci. 51: 725–730.

Boyd NS and Van Acker RC (2004) Seed germination of common weed species as affected by oxygen concentration, light, and osmotic potential. Weed Sci. 52: 589–596.

Cadman CS, Toorop PE, Hilhorst HW and Finch-Savage WE (2006) Gene expression profiles of Arabidopsis Cvi seeds during dormancy cycling indicate a common underlying dormancy control mechanism. The Plant J. 46: 805–22.

Chauhan BS, Gill G and Preston C (2006) Factors affecting seed germination of threehorn bedstraw (*Galium tricornutum*) in Australia. Weed Sci. 54: 471–477.

Chiwocha SD, Cutler AJ, Abrams SR, Ambrose SJ, Yang J, Ross AR and Kermode AR (2005) The etr1-2 mutation in *Arabidopsis thaliana* affects the abscisic acid, auxin, cytokinin and gibberellin metabolic pathways during maintenance of seed dormancy, moist-chilling and germination. The Plant J. 42: 35–48.

Cussans GW, Raudonius S, Brain P and Cumbenworth S (1996) Effects of depth of seed burial and soil aggregate of *Alopecurus myosuroides, Galium aparine, Stellaria media* and wheat. Weed Res. 36: 133–141.

Davis AS, Schutte BJ, Iannuzzi J and Renner KA (2008) Chemical and physical defense of weed seeds in relation to soil seedbank persistence. Weed Sci. 56: 676–684.

Deregibus VA, Casal JJ, Jacobo EJ, Gibson D, Kauffman M and Rodriguez AM (1994) Evidence that heavy grazing may promote the germination of *Lolium multiflorum* seeds via phytochrome-mediated perception of high red/far red ratios. Funct. Ecol. 8: 536–542.

Donohue K, Dorn L, Griffith C, Kim E, Aguilera A, Polisetty CR and Schmitt J (2005) Environmental and genetic influences on the germination of *Arabidopsis thalliana* in the field. Evolution 59: 740–57.

Fenner M and Thompson K (1998) The Ecology of Seeds. 1st ed. Cambridge: Cambridge University Press. 250 p.

Fennimore SA and Foley ME (1998) Genetic and physiological evidence for the role of gibberellic acid in the germination of dormant *Avena fatua* seeds. J. Exp. Botany 49: 89–94.

Finch-Savage WE and Leubner-Metzger G (2006) Seed dormancy and the control of germination. New phytologist 171: 501–23.

Finkelstein R, Reeves W, Ariizumi T and Steber C (2008) Molecular aspects of seed dormancy. Plant Biol. 59: 387.

Footitt S, Douterelo-Soler I, Clay H and Finch-Savage WE (2011) Dormancy cycling in Arabidopsis seeds is controlled by seasonally distinct hormone-signaling pathways. Proc. Natl. Acad. Sci. USA 108: 20236–41.

Forcella F, Arnold RLB, Sanchez R and Ghersa CM (2000) Modeling seedling emergence. Field Crops Res. 67: 123–139.

Froud-Williams RJ, Chancellor RJ and Drennan DSH (1983) Influence of cultivation regime on buried weed seeds in arable cropping systems. J. Applied Ecol. 20: 199–208.

Gardarin A and Colbach N (2015) How much of seed dormancy in weeds can be related to seed traits? Weed Res. 55: 14–25.

Gardarin A, Durr C, Mannino MR, Busset H and Colbach N (2010) Seed mortality in the soil is related to seed coat thickness. Seed Sci. Res. 20: 243–56.

Ghassemian M, Nambara E, Cutler S, Kawaide H, Kamiya Y and McCourt P (2000) Regulation of abscisic acid signaling by the ethylene response pathway in Arabidopsis. The Plant Cell 12: 1117–1126.

Ghorbani R, Seel W and Leifert C (1999) Effects of environmental factors on germination and emergence of *Amaranthus retroflexus*. Weed Sci. 47: 505–510.

Graeber KA, Nakabayashi K, Miatton E, Leubner-Metzger G and Soppe WJ (2012) Molecular mechanisms of seed dormancy. Plant Cell Env. 35: 1769–86.

Grundy AC, Mead A and Burston S (2003) Modelling the emergence response of weed seeds to burial depth: interactions with seed density, weight and shape. J. Applied Ecol. 40: 757–70.

Gulden RH and Shirtliffe SJ (2009) Weed seed banks: biology and management. Prairie Soils and Crops 2: 46–52.

Guo P and Al-Khatib K (2003) Temperature effects on germination and growth of redroot pigweed (*Amaranthus retroflexus*), Palmer amaranth (*A. palmeri*), and common waterhemp (*A. rudis*). Weed Sci. 51: 869–875.

Gutterman Y, Corbineau F and Come D (1992) Interrelated effects of temperature, light and oxygen on *Amaranthus caudatua* L. seed germination. Weed Res. 32: 111–117.

Harrison SK, Regnier EE, Schmoll JT and Harrison JM (2007) Seed size and burial effects on giant ragweed (*Ambrosia trifida*) emergence and seed demise. Weed Sci. 55: 16–22.

Heap I (2016) The international survey of herbicide resistant weeds. Accessed January 15. Available www.weedscience.org.

Hennig L, Stoddart W, Dieterle M, Whitelam G and Schafer E (2002) Phytochrome E controls light-induced germination of *Arabidopsis*. Plant Physiol. 128: 194–200.

Holdsworth MJ, Bentsink L and Soppe WJ (2008) Molecular networks regulating Arabidopsis seed maturation, after-ripening, dormancy and germination. New Phytologist 179: 33–54.

Holm RE (1972) Volatile metabolites controlling weed germination in soil. Plant Physiol. 50: 293–297.

Hou JQ and Simpson GM (1991) Effects of prolonged light on germination of six lines of wild oat (*Avena fatua*). Can. J. Bot. 69: 1414–1417.

Hou JQ and Simpson GM (1993) Germination response to phytochrome depends on specific dormancy states in wild oat (*Avena fatua*). Can. J. Bot. 71: 1528–1532.

Huang X, Schmitt J, Dorn L, Griffith C, Effgen S, Takao S, Koornneef M and Donohue K (2010) The earliest stages of adaptation in an experimental plant population: strong selection on QTLS for seed dormancy. Molecular Ecol. 19: 1335–51.

Imbert E (2002) Ecological consequences and ontogeny of seed heteromorphism. Perspectives in Plant Ecol. Evolution Systematics 5: 13–36.

Imbert E and Ronce O (2001) Phenotypic plasticity for dispersal ability in the seed heteromorphic Crepissancta (Asteraceae). Oikos 93: 126–34.

Insausti P, Soriano A and SaÂnchez RA (1995) Effects of flood-influenced factors on seed germination of *Ambrosia tenuifolia*. Oecologia (Berl.) 103: 127–132.

Jha P, Norsworthy J, Kumar V and Reichard N (2015) Annual changes in temperature and light requirements for *Ipomoea purpurea* seed germination with after-ripening in the field following dispersal. Crop Prot. 67: 84–90.

Jha P, Norsworthy J, Riley M and Bridges Jr W (2010) Annual changes in temperature and light requirements for germination of Palmer amaranth (Amaranthus palmeri) seeds retrieved from soil. Weed Sci. 58: 426–432.

Karssen CM (1970) Photoperiodic induction of dormancy in Chenopodium. Acta Bot. Neerl. 19: 81–94.

Kępczynski J, Bihun M and Kępczynska E (1997) Ethylene involvement in the dormancy of *Amaranthus* seeds. pp. 113–122. In: Kanellis AK, Chang C, Kende H and Grierson D (eds.). Biology and Biotechnology of the Plant Hormone Ethylene. Netherlands: Springer.

Kon KF, Follas GB and James DE (2007) Seed dormancy and germination phenology of grass weeds and implications for their control in cereals. New Zealand Plant Protection 60: 174.

Koornneef M, Bentsink L and Hilhorst H (2002) Seed dormancy and germination. Current Opinion in Plant Biol. 5: 33–36.

Kriticos DJ, Sutherst RW, Brown JR, Adkins SW and Maywald GF (2003) Climate change and the potential distribution of an invasive alien plant: *Acacia nilotica* ssp. *indica* in Australia. J. Applied Ecol. 40: 111–124.

Kucera B, Cohn MA and Leubner-Metzger G (2005) Plant hormone interactions during seed dormancy release and germination. Seed Sci. Res. 15: 281–307.

Le Page-Degivry MT, Garello G and Barthe P (1997) Changes in abscisic acid biosynthesis and catabolism during dormancy breaking in *Fagus sylvatica* embryo. J. Plant Growth Regulation 16: 57–61.

Leishman MR, Wright IJ, Moles AT and Westoby M (2000) The evolutionary ecology of seed size. pp. 31–57. In: Fenner M (ed.). Seeds: The Ecology of Regeneration in Plant Communities. CAB International, Wallingford, UK.

Li Q, Tan J, Li W, Yuan G, Du L, Ma S and Wang J (2015) Effects of environmental factors on seed germination and emergence of Japanese Brome (*Bromus japonicus*). Weed Sci. 63: 641–646.

Li XJ, Zhang MR, Wei SH and Cui HL (2012) Influence of environment factors on seed germination and seedling emergence of yellowtop (*Flaveria bidentis*). Pak. J. Weed Sci. Res. 18: 317–325.

Li YP and Feng YL (2009) Differences in seed morphometric and germination traits of crofton weed (*Eupatorium adenophorum*) from different elevations. Weed Sci. 57: 26–30.

Linkies A and Leubner-Metzger G (2012) Beyond gibberellins and abscisic acid: how ethylene and jasmonates control seed germination. Plant Cell Reports 31: 253–270.

Liu Y, Fang J, Xu F, Chu J, Yan C, Schläppi MR, Wang Y and Chu C (2014) Expression patterns of ABA and GA metabolism genes and hormone levels during rice seed development and imbibition: a comparison of dormant and non-dormant rice cultivars. J. Genet. Genomics 41: 327–338.

López-Granados F and García-Torres L (1999) Longevity of crenate broomrape (*Orobanche crenata*) seed under soil and laboratory conditions. Weed Sci. 47: 161–166.

Matilla AJ (2000) Ethylene in seed formation and germination. Seed Sci. Res. 10: 111–126.

Matilla AJ and Matilla-Vazquez MA (2008) Involvement of ethylene in seed physiology. Plant Sci. 175: 87–97.

Mohler C, Iqbal J, Shen J and DiTommaso A (2016) Effects of water on recovery of weed seedlings following burial. Weed Sci. 64: 285–293.

Moise JA, Han S, Gudynaite-Savitch L, Johnson DA and Miki BLA (2005) Seed coats: structure, development, composition, and biotechnology. *In Vitro* Cell Dev. Biol.-Plant 41: 620–644.

Moles AT, Hodson DW and Webb CJ (2000) Seed size and shape and persistence in the soil in the New Zealand flora. Oikos 89: 541–545.

Nandula VK, Eubank TW, Poston DH, Koger CH and Reddy KN (2006) Factors affecting germination of horseweed (*Conyza canadensis*). Weed Sci. 54: 898–902.

Norsworthy J and Oliveira M (2007) Role of light quality and temperature on pitted morningglory (*Ipomoea lacunosa*) germination with after-ripening. Weed Sci. 55: 111–118.

Pareja RDM and Staniforth W (1985) Seed–soil characteristics in relation to weed seed germination. Weed Sci. 33: 190–195.

Rana N, Sellers BA, Ferrell JA, MacDonald GE, Silveira ML and Vendramini JM (2015) Integrated management techniques for long term control of giant smutgrass (*Sporobolus indicus* var. *pyramidalis*) in bahiagrass pastures. Weed Technol. 29: 570–577.

Rana N, Wilder BJ, Sellers BA, Ferrell JA and MacDonald GE (2012) Effects of environmental factors on seed germination and emergence of smutgrass (*Sporobolus indicus*) varieties. Weed Sci. 60: 558–563.

Roberts HA and Feast PM (1973) Emergence and longevity of seeds of annual weeds in cultivated and undisturbed soil. The J. of Appl. Ecol. 10: 133–143.

Roberts HA and Potter ME (1980) Emergence patterns of weed seedlings in relation to cultivation and rainfall. Weed Res. 20: 377–386.

Rodríguez-Gacio MDC, Matilla-Vázquez MA and Matilla AJ (2009) Seed dormancy and ABA signaling: the breakthrough goes on. Plant Signaling and Behavior 4: 1035–1048.

Romo JT and Haferkamp MR (1987) Forage Kochia germination response to temperature, water stress, and specific ions. 79: 27–30.

Russell EJ and Appleyard A (1915) The atmosphere of the soil: its composition and the causes of variation. J. Agr. Sci. 7: 1–48.

Saatkamp A (2009) Population dynamics and functional traits of annual plants—A comparative study on how rare and common arable weeds persist in agroecosystems. Ph.D. Dissertation. Regensberg, Germany: University of Regensburg, Regensberg, Germany. 220 p.

Scopel AL, Ballare CL and Radosevich SR (1994) Photostimulation of seed germination during soil tillage. New Phytol. 126: 145–152.

Scopel AL, Ballare CL and Sanchez RA (1991) Induction of extreme light sensitivity in buried weed seeds and its role in the perception of soil cultivation. Plant Cell Environ. 14: 501–508.

Shinomura T, Nagatani A, Chory J and Furuya M (1994) The induction of seed germination in *Arabidopsis thaliana* is regulated principally by phytochrome B and secondarily by phytochrome A. Plant Physiol. 104: 363–371.

Souza FH and Marcos-Filho JU (2001) The seed coat as a modulator of seed-environment relationships in Fabaceae. Brazilian J. Botany 24: 365–75.

Steber CM and McCourt P (2001) A role for brassinosteroids in germination in Arabidopsis. Plant Physiol. 125: 763–769.

Sun T and Gubler F (2004) Molecular mechanism of gibberellin signaling in plants. Annu. Rev. Plant Biol. 55: 197–223.

Tanveer A, Tasneem M, Khaliq A, Javaid MM and Chaudhry MN (2013) Influence of seed size and ecological factors on the germination and emergence of field bindweed (*Convolvulus arvensis*). Planta Daninha 31: 39–51.

Taylorson RB (1972) Light sensitivity of buried seeds. Weed Sci. 20: 417–425.

Teasdale JR and Mohler CL (1993) Light transmittance, soil temperature, and soil moisture under residue of hairy vetch and rye. Agron. J. 85: 673–680.

Thompson K, Band SR and Hodgson JG (1993) Seed size and shape predict persistence in soil. Functional Ecol. 7: 236–241.

Topp GC, Dow B, Edwards M, Gregorich EG, Curnoe WE and Cook FJ (2000) Oxygen measurements in the root zone facilitated by TDR. Can. J. Soil Sci. 80: 33–41.

Tozzi E, Beckie H, Weiss R, Gonzalez-Andujar J, Storkey J, Cici S and Van Acker R (2013) Seed germination response to temperature for a range of international populations of *Conyza canadensis*. Weed Res. 54: 178–185.

Traba J, Azcárate FM and Peco B (2004) From what depth do seeds emerge? A soil seed bank experiment with Mediterranean grassland species. Seed Sci. Res. 14: 297–304.

Treharne K (1989) The implications of the greenhouse effect for fertilizers and agrochemicals. pp. 67–78. *In*: Bennet RM (ed.). The Greenhouse Effect and UK Agriculture. Reading, UK: University of Reading Press CAS Paper 19.

Van Mölken T, Jorritsma-Wienk LD, Van Hoek PH and Kroon HD (2005) Only seed size matters for germination in different populations of the dimorphic *Tragopogon pratensis* subsp. *pratensis* (Asteraceae). Am. J. Botany 92: 432–437.

Vidotto F, De Palo F and Ferrero A (2013) Effect of short-duration high temperature on weed seed germination. Ann. Appl. Biol. 163: 454–465.

Weber H, Borisjuk L and Wobus U (2005) Molecular physiology of legume seed development. Annu. Rev. Plant Biol. 56: 253–279.

Wesson G and Wareing PF (1967) Light requirements of buried seeds. Nature 213: 600–601.

Wesson G and Wareing PF (1969) Buried seeds and light. J. Exp. Bot. 20: 414–425.

Westonby M (1998) A leaf-height-seed (LHS) plant ecology strategy scheme. Plant and Soil 199: 213–227.

Willenborg CJ, Wildeman JC, Miller AK, Rossnagel BG and Shirtliffe SJ (2005) Oat germination characteristics differ among genotypes, seed sizes, and osmotic potentials. Crop Sci. 45: 2023–2029.

Woolley JT and Stoller EW (1978) Light penetration and light-induced seed germination in soil. Plant Physiol. 61: 597–600.

WSSA Weeds Surveys (2016) 2015 Survey—Baseline survey for 26 different crop, non-crop, aquatic, and natural areas (http://wssa.net/wssa/weed/surveys/).

Yamauchi Y, Ogawa M, Kuwahara A, Hanada A, Kamiya Y and Yamaguchi S (2004) Activation of gibberellin biosynthesis and response pathways by low temperature during imbibition of *Arabidopsis thaliana* seeds. The Plant Cell 16: 367–378.

Yamauchi Y, Takeda-Kamiya N, Hanada A, Ogawa M, Kuwahara A, Seo M, Kamiya Y and Yamaguchi S (2007) Contribution of gibberellin deactivation by AtGA2ox2 to the suppression of germination of dark-imbibed *Arabidopsis thaliana* seeds. Plant Cell Physiol. 48: 555–561.

Chapter 4

A New Approach to Weed Control in Cropping Systems

Michael Walsh[1,]* and *Bhagirath Singh Chauhan*[2]

Introduction

Worldwide, infestations of weeds in cropping systems are ubiquitous and an annual threat to productivity, especially in the major crops (wheat [*Triticum aestivum* L.], rice [*Oryza sativa* L.], soybean [*Glycine max* (L.) Merr.], corn [*Zea mays* L.], canola [*Brassica napus* L.], etc.). This threat is significant and must be minimized to maximize crop productivity and thus global food supply/security. Currently, herbicides are the dominant technology used against infesting crop weed populations. The many advantages of herbicides over other forms of weed control have resulted in almost exclusive reliance on this technology in cropping systems throughout the worlds' cropping regions. However, the exposure of huge weed populations over vast areas to strong herbicide selections has inevitably resulted in the evolution of herbicide-resistant weed populations (Heap 2015).

Dramatic and widespread herbicide resistance in weed populations is just another example of evolution in weed populations exposed to a control tactic or a cropping system selection pressure. There are numerous instances of weed adaptations occurring in response to changes in cropping system practices (Gould 1991). An often repeated example is the evolution of dormancy and specific germination requirements in response to exposure to frequent tillage treatments that once were standard practice for weed

---

[1] Faculty of Agriculture and Environment, School of Life and Environmental Sciences, University of Sydney, 12656 Newell Highway, Narrabri, NSW 2390, Australia.
[2] The Centre for Plant Science, Queensland Alliance for Agriculture and Food Innovation (QAAFI), University of Queensland, Gatton, Queensland 4343, Australia.
* Corresponding author

control and seed bed preparation (Jana and Thai 1987; Naylor and Jana 1976). More recent examples of adaptation have occurred following the introduction of conservation cropping systems where weed populations have evolved to persist in cropping systems where there is much reduced soil disturbance (Fleet and Gill 2012; Kleemann and Gill 2006). A further example is evidenced by induced dormancy variations within a golf course population of *Poa annua* L. in response to mowing treatments (Wu et al. 1987). In this instance plants growing in the golf course 'rough' that received less frequent mowing produced seeds with a higher level of dormancy than plants growing on the fairway or green where mowing was more frequent. Crop mimicry has been documented as a weed populations' evolutionary response to hand weeding where weeds have adapted to this selection pressure by mimicking crop colour and architecture (Baker 1974; Londo and Schaal 2007). Further, the uniform presence of populations of a weed species in cropping systems spread across a diverse range of environments and production practices is clear evidence of adaptation and the evolutionary potential of a weed species. Many of the dominant weeds (*Alopecurus myosuroides* Huds., *Avena* spp., *Bromus* spp., *Kochia scoparia* L., *Lolium rigidum* Gaud., *Phalaris* spp., and *Raphanus raphanistrum* L.) of global cropping regions are established and problematic in production systems conducted across a diverse range of environments with highly varied production practices (Collavo et al. 2011; Kercher and Conner 1996; Kloot 1983; Kon and Blacklow 1989; Menchari et al. 2007; Mengistu and Messersmith 2002). These weed species cannot persist in these highly variable environments without the occurrence of evolution and adaptation.

The dominant weeds of global cropping systems are all characteristically genetically diverse with breeding systems that encourage rapid adaptation. The dramatic evolutionary consequences of this genetic diversity are clearly evident in the rapid and frequent evolution of herbicide resistance in very many biotypes of these species (Heap 2015). As highlighted in Neve et al. (2009), there is little doubt that weed populations, given the opportunity, will evolve or adapt to resist any and all forms of weed control. Whether resistance/adaptation occurs through the selection of initially rare major gene traits (i.e., 1×10^{-4} to 1×10^{-9}) (Jasieniuk et al. 1996; Preston and Powles 2002) or from the accumulation of quantitative traits (Délye 2013; Neve and Powles 2005), population size governs this evolutionary process. Resistance or adaptation to specific weed control strategies, regardless of mechanism or type, will evolve more frequently and rapidly in large populations persistently exposed to selection (Diggle and Neve 2001; Jasieniuk et al. 1996; Jordan and Jannink 1997). Restricting the size of weed populations is critical in delaying, if not preventing, these evolutionary processes in genetically diverse weed populations and the inevitable loss of effective weed control strategies. Given this diversity then it is only under very low (e.g., < 1.0 plants/m^2) weed population scenarios that the preservation of weed control techniques can be achieved. Thus, in the face of this threat the focus for producers now needs to be 'zero tolerance' of in-crop weeds. However, as history has shown that a single control strategy cannot be relied upon and a diversity of weed control strategies is needed to achieve acceptably low weed population densities.

Weed Control within the Conservation Cropping Systems Context

Current and future cropping practices will more frequently lie within the conservation production approach, based on reduced tillage and crop residue retention. The increasing need to improve soil structure, retain nutrients, and conserve soil moisture will continue to drive the widespread adoption of conservation cropping practices (FAO 2015; Kassam et al. 2012; Llewellyn et al. 2012). These systems are now proven to be more robust and sustainable, providing some stability in spite of highly variable climatic conditions. Thus, even though conservation cropping has led to herbicide reliance and subsequently herbicide resistance, this approach has provided substantial gains in productivity and reverting to less conservative systems for the sake of weed control is no longer an option. However, practices such as stubble burning and tillage can continue to be used for weed control in conservation cropping systems but their use now must be strategic, and not routine as previously practised. Herbicides are likely to remain the focus of crop weed control programs but the challenge is to identify other suitable weed control technologies that can also be routinely used alongside herbicides.

Herbicides: resistance and future resources

Regardless of the ever increasing frequency of herbicide-resistant weed biotypes (Heap 2015), herbicides remain the most effective form of weed control in global cropping systems. However, the herbicide industry has shifted focus away from the search for new molecules to the development of gene traits conferring herbicide resistance in crop species (Green 2014). This has resulted in a dramatic slowing of the introduction of new herbicides (Duke 2012) which, combined with regulatory removal of some older herbicides (Chauvel et al. 2012), means that herbicide resources continue to diminish. However, no other currently available form of weed control provides similar efficacy combined with excellent crop safety and flexibility of use across a diverse range of conditions and situations. In addition, herbicides are by far the most suited form of weed control for conservation cropping systems. Thus, the need for new herbicides is high and will remain so for the foreseeable future but it remains to be seen if this is enough of a stimulus for herbicide discovery companies to continue their search for new molecules.

The threat of diminishing herbicide resources is the driving force in the search for alternate weed control technologies that can be used to sustain the ongoing use of the current herbicide resources (Walsh et al. 2013). Thus, in the short term, until new herbicide molecules are developed, the current herbicide resources will continue to be relied on and herbicide resistance evolution will continue to be a major threat to future crop weed control. The inclusion of additional control techniques in weed management programs is essential if growers are going to prevent losing forever valuable herbicide resources.

Additional Weed Control Technologies

The conservation of herbicide resources as well as all other weed control tactics is reliant on the inclusion of additional weed control technologies into weed management programs. However, additional weed control techniques can only be included for routine use in these programs if they do not interfere with the current conservation crop production systems. The following discusses the currently available weed control options that fit with this criteria.

Competitive crops

The use of agronomic practices to enhance the competitiveness of crops is a simple cost-effective approach that can be used to complement other in-crop weed control practices. Enhancing crop competiveness aims at maximising crop resource utilisation to the detriment of weed populations. In the absence of control, weeds compete for the essential primary resources of nutrients, water, and sunlight, reducing their availability for wheat crops (Roush and Radosevich 1985). Much of the focus of crop competition research has been on investigating how enhanced crop competition can reduce the impact of in-crop weed populations on crop growth and yield. As reviewed in Zimdahl (2007), there is substantial evidence demonstrating the efficacy of many agronomic techniques that provide crops with a competitive advantage over infesting weed populations, inevitably leading to increased grain yields. However, in our search for additional weed control strategies we also need to consider crop competition as an effective form of weed control.

Predominantly, the opportunities for enhancing the competitive effects of crops on weed populations are implemented at crop seeding. Crop cultivar, seed size, seeding rate, row spacing, row orientation, and fertiliser placement can all be adjusted to ensure establishing crop seedlings have a competitive advantage over weeds (Andrew et al. 2015; Blackshaw 2004; Borger et al. 2009; Lemerle et al. 2001; Lemerle et al. 2004; Lutman et al. 2013; Yenish and Young 2004; Zerner et al. 2008). Most studies involving weed-competitive cultivars have focussed on aboveground traits (e.g., leaf area index, plant biomass, tillering capacity, height). Very little attention has been given to the role of root competition for nutrients and water. Breeders and physiologists need to include root traits when evaluating weed-competitive cultivars, especially in dry land agriculture. Competition studies have focussed on measuring weed biomass as an indicator of crop competition effects, unfortunately, these studies in general have neglected the assessment of the impact on weed fecundity, arguably the most significant assessment of the impact of crop competition. As reviewed in Norris (2007), some studies have highlighted reduced fecundity due to crop competition. For example increasing corn plant densities from 25000 to 50000 plants ha^{-1} reduced *Amaranthus retroflexus* L. seed production from 54,000 to 6,200 seed plant^{-1} (McLachlan et al. 1995). Even weakly competitive crops such as kidney beans (*Phaseolus vulgaris* L.) can reduce the seed production of *Abutilon theophrasti* Medik. by approximately 90%. Crop competition can substantially impact on fecundity of some of the world's major weeds, e.g., *A. myosuroides* (Lutman et al. 2013), *R. raphanistrum* (Walsh and Minkey 2006), *L. rigidum* (Pedersen et al. 2007), *A. palmerii* (Webster and Grey 2015), etc. reducing seedbank inputs by > 50% and, therefore, needs to be considered in planning weed management programs.

Enhanced crop competition cannot be considered a standalone weed control treatment but when combined with other weed control practices, the additional impact on weed populations can be critical towards achieving weed control. For example, enhanced wheat crop competition will routinely increase the efficacy of selective herbicide treatments in controlling crop-weed populations (Kim et al. 2002). Importantly, this competition can lead to the control of weed populations that are resistant to the applied herbicide. For example, a 2,4-D resistant *R. raphanistrum* population was controlled when 2,4-D was applied at the recommended rate to resistant plants present within a competitive wheat crop (Walsh et al. 2009). As well as complementing herbicide activity, enhanced wheat competition will likely improve the efficacy of HWSC strategies. Annual weed species infesting global wheat production systems are typically not shade tolerant (Gommers et al. 2013) and as indicated from competition studies, grow poorly when shaded (Zerner et al. 2008). When competing with wheat for light the likely response for shade intolerant weed species is a more upright growth habit (Morgan et al. 2002; Vandenbussche et al. 2005). This erect growth habit will undoubtedly lead to higher proportions of total seed production being located above harvester cutting height and increasing subsequent exposure to harvest weed seed control (HWSC) methods. Clearly then, the combined benefits of higher yield potential and enhanced weed control ensure that agronomic weed management should be standard practice throughout global wheat production systems.

Cover crops and mulching

Any substance or material applied on the soil surface to avoid the exposure of soil to external factors may be regarded as mulch. Mulches may be of several types, depending upon the nature of material they consist of, modes of application, objectives, and functions (Bond and Grundy 2001). Besides multiple benefits, mulching has great implications for weed control when either in the form of live cover crops or as surface retained crop residues. In addition to suppressing weeds, cover crops and mulches reduce soil erosion, enhance soil fertility, and can conserve soil moisture.

A number of legumes (e.g., alfalfa, cowpea, clover, lupins, and sunhemp) and non-legumes (e.g., rye and buckwheat) are commonly grown as cover crops to suppress and smother weeds by exhibiting allelopathic effects and decreasing light transmittance to the soil surface (Chauhan et al. 2012). For example, use of leguminous crops as cover crop reduced the population of *Echinochloa crus-galli* (L.) P. Beauv. and use of barley as a cover crop has been shown to suppress many weeds in soybean (Caamal-Maldonado et al. 2001; Kobayashi et al. 2004). Similarly, rye residues were found to reduce the emergence of *Setaria viridis* (L.) P. Beauv., *Amaranthus retroflexus* L., and *Portulaca oleracea* L. by 80, 90, and 100%, respectively (Putnam and DeFrank 1983). Compared with the no residue treatment, the residue of Russian vetch and rye reduced total weed density by more than 75% (Mohler and Teasdale 1993). Similarly, small grain crop residues reduced the densities of *Ipomoea* spp., *Sida spinosa* L., and *Cassia obtusifolia* L. (Liebl and Worsham 1983). Sorghum mulch at the rate of 10–15 t ha^{-1} provided effective control of *Cyperus rotundus* L. and *Trianthema portulacastrum* L. (Cheema et al. 2004).

In addition to reducing weed emergence and weed growth, the use of crop residues as mulch can also delay weed emergence (Chauhan and Abugho 2013). This delayed emergence may have implications for weed management. Late emerging weed seedlings are likely to experience greater competition from the crop than early emerging weed seedlings, which may result in lower weed seed production and less crop yield loss. Crop residues suppress weeds early in the growing season, but the use of only residue as mulch may not provide complete control of weeds. Therefore, it is necessary to integrate herbicide use or other strategies with residue retention to achieve season-long weed control (Chauhan 2012; Chauhan et al. 2012; Teasdale 1993).

Targeting Seed Production

Crop topping

There is now worldwide recognition that the in-crop annual weed populations result from the previous season's seed production. This realisation has stimulated a focus on minimising weed seed production and inputs to the seedbank. In Australia, this focus on weed seed production became a major driver in development of techniques aimed at dealing with weed populations (McGowan 1970; Pearce and Holmes 1976; Reeves and Smith 1975). Spray-topping is one such Australian developed technique where a non-selective herbicide (paraquat or glyphosate) is applied over a crop to target weed seed production. In situations where the crop has matured while weed seeds remain immature there is a time period, 'window of opportunity' during which weed seeds can effectively be targeted without incurring crop yield loss. In favourable situations, weed seed production is frequently reduced by up to 90% for a range of weed species including *L. rigidum* (Gill and Holmes 1997; Steadman et al. 2006), *R. raphanistrum* (Walsh and Powles 2009), *Vulpia bromoides* (L.) Gray (Leys et al. 1991), *Hordeum glaucum* Steud. (Powles 1986), and *Bromus* spp. (Dowling and Nicol 1993). The reproductive phase of many annual weeds is strongly influenced by seasonal conditions that often prolong the period of seed maturation (Aitken 1966), resulting in an overlap with the more defined maturation phase of crop species. A consequence of this is that there is little or no 'window of opportunity' for the application of effective crop-topping treatments. Frequently then, the timing of crop-topping treatments is a trade-off between acceptable levels of yield loss and effective weed seed control. Despite this crop-topping is routinely used in non-cereal crops (e.g., lupins, peas), some yield loss is often tolerated in preference to maximising weed seed control prior to the next wheat crop.

Harvest weed seed control

The biological attribute of seed retention at maturity in annual weed species means that intact seed heads remain at crop maturity enabling the weed seeds to be collected (harvested) during grain crop harvest. For example, in field crops a large proportion (up to 80%) of total *L. rigidum* seed production can be collected during a typical commercial grain harvest (Walsh and Powles 2007). These weed seeds enter and are processed by the grain harvester and then immediately exit the grain harvester, usually in the chaff fraction and, ironically are evenly redistributed across the crop field to become future weed problems. Thus, grain crop harvest presents an opportunity to target weed seed production and exploit high weed seed retention by targeting the collected weed seeds as they pass through the harvester to minimise inputs to the weed seed bank. In Australia, HWSC systems have been developed to target and destroy weed seeds during commercial grain crop harvest preventing inputs into the seed bank (Walsh and Powles 2007). With the major portion (> 95% *L. rigidum*) of weed seeds collected during harvest exiting in the chaff fraction (Walsh and Powles 2007), this material then was the focus for HWSC techniques. The first system developed was the chaff cart which originated in Canada where known as chaff wagons, was a trailing cart attached to the harvest that collected the chaff material during harvest (Olfert et al. 1991). In Canada, this material was collected specifically for livestock feed; however, in Australia, the focus was on targeting the collected weed seeds. This relatively simple HWSC system consists of a chaff collection and transfer mechanism, attached to a grain harvester that delivers the weed seed bearing chaff fraction into a bulk collection bin, usually a trailing cart. Chaff cart collection systems have been shown to achieve the collection and removal of high proportions of seed from *L. rigidum, R. raphanistrum* (Walsh and Powles 2007) and *Avena* spp. (Shirtliffe and Entz 2005). Because of the large volume of material, the collected chaff is typically dumped in chaff heaps, in lines across fields in preparation for subsequent burning to ensure weed seed destruction. As identified earlier, chaff material is a valuable feed source and can be grazed *in-situ* by livestock and, in some instances, collected for use in livestock feed-lots. The adoption in Australia of chaff cart collection systems, although limited, has proven to be very effective for intensive grain producers faced with major herbicide-resistant weed problems.

The most widely adopted HWSC system in use in Australia is narrow windrow burning where a grain harvester mounted chute concentrates all of the chaff and straw residue into a narrow windrow (500–600 mm). These chaff and straw windrows are subsequently burnt, without burning the entire crop field. The concentration of chaff and straw residues increases the duration and temperature of burning treatment ensuring weed seed destruction. Weed seed kill levels of 99% for both *L. rigidum* and *R. raphanistrum* have been recorded from the burning of wheat, canola, and lupin narrow windrows (Walsh and Powles 2007). The simplicity and low cost of narrow-windrow burning has resulted in the adoption of this technique by an estimated 30% of Australian crop producers.

The Bale Direct System consists of a large square baler directly attached to the harvester that bales chaff and straw residues during grain crop harvest. This system was developed as a method for effectively collecting harvest residues for subsequent use as livestock feed or bedding. Studies have determined that very high proportions (e.g., 95% *L. rigidum*) of weed seeds are also removed (Walsh et al. 2013; Walsh and Powles 2007). The bale direct system is highly suited for use in areas where high yielding crops produce large amounts of stubble residues that need to be managed before seeding of the next crop. At present, the availability of suitable livestock markets for the use of baled material as a feed source has limited the adoption of this system. However, in the future, there will potentially be opportunities for using the baled material for other uses such as energy generation.

An innovation by a Western Australian grower Ray Harrington, the Harrington Seed Destructor (HSD), represents a new weed control tool that is now commercially available to growers. The HSD is a trailer mounted, cage mill based chaff processing system, incorporating chaff and straw deliver systems. Chaff is delivered from the rear of the harvester to the cage mill which processes this material sufficiently to destroy the contained weed seeds. This system processes chaff material sufficiently to destroy greater than 95% of the seed of major annual weed species *L. rigidum, R. raphanistrum, Avena* spp., and *Bromus* spp. (Walsh et al. 2012).

Although the introduction of HWSC systems into Australian cropping was promulgated by excessive frequencies of herbicide-resistant weed populations when included as an additional strategy in weed management programs, HWSC systems can facilitate very low weed population densities. A 14-year Australian study found that the inclusion of HWSC in herbicide-based weed control programs resulted in substantially lower *L. rigidum* densities. In 25 fields where *L. rigidum* populations were problematic (> 50 plants m^{-2}) the combination of herbicide use and HWSC led to average *L. rigidum* densities being reduced to < 1.0 plant m^{-2} compared to 5 plants m^{-2} when herbicides alone were used (Walsh et al. 2013). Remarkably these additional reductions in weed numbers occurred when HWSC systems were used in just one year in three.

Identifying New Weed Control Opportunities

Effective weed management in crop production systems are based on knowledge of the biology of the targeted weed species as they occur in these systems. All weed control techniques require some basic understanding of weed biology for their effective implementation (Bhowmik 1997; Van Acker 2009). Even the introduction of herbicides, likely the most effective weed control tool we will ever have, has not diminished the need for weed biology. In fact the subsequent widespread occurrence of herbicide resistance evolution has resulted in our current knowledge of genetics and evolutionary adaptation in weed species (Jasieniuk et al. 1996; Powles and Yu 2010). In hindsight, the dramatic impacts of herbicide resistance may not have occurred if there had been a better understanding of selection pressure and the genetic potential in weed populations at the time of herbicide introduction. However, a harsh lesson from herbicide resistance is the need for additional new weed control technologies to be adopted if weed management programs are going to be successful (Walsh et al. 2013).

An Occasional Intervention

The intervention approach is a new strategy that needs to be included in weed management programs where drastic weed control action is taken when weed population densities reach a pre-determined critically elevated density (e.g., > 1.0 plant/m^2). In all weed management programs, there will be instances when weed densities increase to a level that places undue pressure on the sustainability of weed control practices. The greatest influence on weed control efficacy is climate and there are a wide range of seasonal conditions that directly influence the efficacy of weed control practices (e.g., drought, waterlogging, frost, high temperatures, etc.). Therefore, even the very best routine weed management programs are not perfect and there will be occasions when additional weed control is needed. Because the threat of resistance evolution to all weed control practices increases with increasing weed densities then a major, disruptive weed control tactic is required that quickly delivers substantially lower weed numbers. Only when a lower weed threshold (e.g., < 1.0 plant/m^2) is reached should the regular weed management program be resumed. In these instances, rather than relying completely on routine control treatments, disruptive control techniques should be used to dramatically reduce large crop weed populations.

Hay, silage and manure crops

Excessively high weed populations and the absence of effective in-crop herbicide treatments can force growers to move away from continuous cropping for one or more years to enable the use of more vigorous approaches to reduce weed numbers. In Australian crop production systems, techniques such as hay, silage (Gill and Holmes 1997) or manure crops (Flower et al. 2012) have been shown to dramatically reduce annual ryegrass populations, often within one season to quickly allow the resumption of continuous cropping. Similarly, implementing a season long fallow phase, provides the opportunity to dramatically reduce weed populations, typically through herbicide use, as well as to conserve soil moisture and provide a disease break (Matthews et al. 1996; Reeves and Smith 1975). Although, these approaches all require an interruption to continuous cropping and frequently lost earning potential, they become necessary when weed numbers are allowed to escalate to a point where crop production is no longer viable.

Strategic tillage

Initially, tillage was used primarily to improve conditions for crop establishment and weed control. However, the advent and successful adoption of no-till systems incorporating chemical weed control has clearly demonstrated that tillage is unnecessary for weed control (Zimdahl 2013). The greater reliance on herbicides, however, can increase the problem of herbicide-resistant weeds in no-till systems. In Australia, for example, *L. rigidum*, *Sonchus oleraceus* L., *R. raphanistrum*, *Echinochloa colona* (L.) Link., *Conyza bonariensis* (L.) Cronq., and *Urochloa panicoides* P. Beauv. have already evolved resistance to glyphosate (Heap 2015). However, despite the risk of evolution of herbicide resistance, these highly productive no-till cropping systems need to be sustained. Therefore, strategic tillage has been receiving great attention among researchers and farmers in several countries (Dang et al. 2015; Melander et al. 2015; Renton and Flower 2015).

Strategic tillage may include an occasional tillage operation of a whole field or a tillage operation for targeting individual weeds or weed patches. In the northern region of Australia, lower densities of *C. bonariensis*, *Raphanus raphanistrum*, *Rapistrum rugosum* (L.) All., and *Avena fatua* L. were reported in the first year following a strategic chisel tillage operation (Crawford et al. 2015). Similarly, another study in Queensland, Australia, reported 61–90% reduced emergence of *Chloris virgata* Sw., *Chloris truncata* R. Br., and *C. bonariensis* after occasional tillage with harrow, gyral, and offset discs compared to a no-till system (McLean et al. 2012).

Recently, mouldboard ploughing has been re-introduced where soil inversion buries the shallow weed seed banks established under long-term conservation cropping systems to a depth from which there is no emergence (> 30 cm) (Code and Donaldson 1996; Reeves and Smith 1975). Prior to the widespread adoption of conservation cropping practices, mouldboard ploughing was for decades routinely used for weed control across the worlds' cropping regions (Cirujeda and Taberner 2009; Lutman et al. 2013; Mas and Verdú 2003; Ozpinar 2006). Now strategic mouldboard ploughing is being used as an effective weed control practice to target weed seed banks in conservation wheat production systems. An occasional tillage of the whole field can be a useful weed control technique and when used sparingly, the positive effects of no-till systems on soil conditions can be retained (Dang et al. 2015).

Although disruptive weed control practices can complicate simple intensive cropping programs, their tactical use reduces the selection pressure on routinely used in-crop control practices. The strategic disruptive weed control, although a major interference in crop production, reduces the selection pressure on routine control practices with the aim of preserving the use of these strategies for the long term.

Site-Specific Weed Management

The opportunity for the introduction of a wide range of novel control tactics and substantial cost savings are driving the future for weed management towards site-specific weed control. Although it is only after very low weed densities (≈ 1 plant/10 m^2) are achieved that site-specific weed management becomes logistically and financially feasible. This approach relies on the detection of weeds and weed patches where the current commercially available options are based on spectral reflectance that can reasonably accurately detect green leaf material against a soil or stubble background (Scotford and Miller 2005). These systems cannot discriminate between crops and weeds but are useful for controlling weeds in non-crop areas such as fallows.

Given the potential cost savings of site-specific weed control versus whole field weed treatments there is now significant effort being focussed on the development of weed detection and mapping systems for in-crop use. The capacity to accurately detect and map low density weed populations within a crop creates the opportunity for the use of range of control tactics that are currently not available. In low weed density situations, because of the small areas involved, and therefore the reduced impact on crop yields, detected weeds can be aggressively targeted. For example, non-selective herbicides, tillage treatments, even hand weeding all become viable options in these situations. Additionally, the ability to strategically target low density weed populations creates the potential for the introduction of more novel weed control technologies such as electrocution (Vigneault et al. 1990), flaming (Bond et al. 2007; Hoyle et al. 2012), microwaves (Brodie et al. 2012), infrared (Ascard 1998) and lasers (Marx et al. 2012). These approaches that currently are not considered due to high energy costs, suddenly become viable for site-specific weed management. Therefore, motivation for the development of weed mapping and identification systems is further driven by the opportunity for additional weed control techniques.

There are a range of approaches being undertaken in the development of weed identification and mapping ranging from vehicle mounted to UAV and even satellite based. However, it is likely that no one single system will suit all situations and production practices. Recently in South Australia, researchers developed a machine vision system which can detect the weeds both in fallows as well as in the inter-row of crops (Liu et al. 2014). The system included three parts: image acquisition, green plant detection, and the inter-row weed detection. Such vision systems can be used to manage herbicide-resistant weeds using strategic tillage. A recent study in Denmark compared the weeding performance of an 'intelligent' camera-based mechanical weeding machine with 'non-intelligent' tool such as a weed harrow, finger and torsion weeders (Melander et al. 2015). Results showed that intelligent weeding capable of cultivating the soil close to vegetable crop plants can provide acceptable weed control without any need for subsequent manual weeding. Non-intelligent tools are non-discriminatory and treat both crops and weeds equally when passing through the intra-row area.

Development and use of Nano-Herbicides

Herbicide efficacy can be reduced due to reduced uptake by weeds or through the applied formulation being unstable and rapidly degraded in the environment before uptake can occur. Herbicide uptake is influenced by several factors, including plant water status and the kind of leaf surface. There is a need to increase the availability and flexibility of control of herbicide-resistant weeds, including new herbicide uses. Herbicide efficacy can potentially be improved through the use of nanotechnology.

Herbicide molecules at nanoscale, called nanoherbicides, have some properties which are different from what they have on a macro scale, enabling their unique applications. Nanoherbicides are 1 to 100 nm in particle size, which means that they are 2,000 to 5,000 times smaller than conventional herbicides. The reduction in particle size enhances surface area of the herbicide molecules, resulting in increased efficacy, uptake by the plant, and solubility in the spray tank. Therefore, the use of such herbicides can help manage weeds in modern agriculture, without leaving toxic residues in the soil and environment (Ali et al. 2014; Pérez-de-Luque and Rubiales 2009). Nanoherbicides could prove very useful in conservation agriculture systems, in which the retention of crop residue on the soil surface can reduce herbicide efficacy, especially for soil active herbicides. Potentially these herbicides will also control weeds that have become resistant to conventional herbicides (Ali et al. 2014).

Conclusion

The future of sustainable weed control is reliant on the attainment of very low weed densities in global cropping systems. It is only under a regime of low densities (e.g., 1.0 plant/10 m^2) that the use of routine weed control practices can be sustained due to the diminished threat of resistance evolution. The problematic annual weeds infesting global cropping systems have clearly demonstrated their propensity for herbicide resistance evolution. The frequent and rapid occurrence of herbicide resistance in these genetically diverse species highlights their potential to evolve resistance to all weed control practices and programs. Therefore, the sustainable management of these weed species is reliant on removing their evolutionary potential by decreasing genetic diversity. The simplest feasible approach for this is to reduce population densities towards diminishing the selection pressure. The mantra now for weed managers is to "drive problematic weed populations towards zero".

LITERATURE CITED

Aitken Y (1966) The flowering responses of crop and pasture species in Australia I. Factors affecting development in the field of *Lolium* species (*L. rigidum* Gaud., *L. perenne* L., *L. multiflorum* Lam.). Aust. J. Agric. Res. 17: 821–839.

Ali MA, Rehman I, Iqbal A, Din S, Rao AQ, Latif A, Samiullah TR, Azam S and Husnain T (2014) Nanotechnology, a new fronteir in Agriculture. Advancements in Life Science 1: 129–138.

Andrew IKS, Storkey J and Sparkes DL (2015) A review of the potential for competitive cereal cultivars as a tool in integrated weed management. Weed Res. 55: 239–248.

Ascard (1998) Comparison of flaming and infrared radiation techniques for thermal weed control. Weed Res. 38: 69–76.

Baker HG (1974) The evolution of weeds. Annual Review of Ecology & Systematics 5: 1–24.

Bhowmik PC (1997) Weed biology: Importance to weed management. Weed Sci. 45: 349–356.

Blackshaw RE (2004) Application method of nitrogen fertilizer affects weed growth and competition with winter wheat. Weed Biology & Management 4: 103–113.

Bond W, Davies G and Turner RJ (2007) A review of thermal weed control. Coventry, UK: Ryton Organic Gardens: HDRA.

Bond W and Grundy AC (2001) Non-chemical weed management in organic farming systems. Weed Research 41: 383–405.

Borger CPD, Hashem A and Pathan S (2009) Manipulating crop row orientation to suppress weeds and increase crop yield. Weed Sci. 58: 174–178.

Brodie G, Ryan C and Lancaster C (2012) Microwave technologies as part of an integrated weed management strategy: A review. International Journal of Agronomy 2012: 14.

Caamal-Maldonado JA, Jimenez-Osorino JI, Barragan AT and Anaya AL (2001) The use of allelopathic legume cover and mulch species for weed control in cropping systems. Agronomy Journal 93: 27–36.

Chauhan BS (2012) Weed ecology and weed management strategies for dry-seeded rice in Asia. Weed Technology 26: 1–13.

Chauhan BS and Abugho SB (2013) Effect of crop residue on seedling emergence and growth of selected weed species in a sprinkler-irrigated zero-till dry-seeded rice system. Weed Science 61: 403–409.

Chauhan BS, Singh RG and Mahajan G (2012) Ecology and management of weeds under conservation agriculture: A review. Crop Protection 38: 57–65.

Chauvel B, Guillemin J-P, Gasquez J and Gauvrit C (2012) History of chemical weeding from 1944 to 2011 in France: Changes and evolution of herbicide molecules. Crop Protect. 42: 320–326.

Cheema ZA, Khaliq A and Saeed S (2004) Weed control in maize (*Zea mays* L.) through sorghum allelopathy. Journal of Sustainable Agriculture 23: 73–86.

Cirujeda A and Taberner A (2009) Cultural control of herbicide-resistant *Lolium rigidum* Gaud. populations in winter cereal in Northeastern Spain. Spanish Journal of Agricultural Research 7: 146–154.

Code GR and Donaldson TW (1996) Effect of cultivation, sowing methods and herbicides on wild radish populations in wheat crops. Aust. J. Exp. Agric. 36: 437–442.

Collavo A, Panozzo S, Lucchesi G, Scarabel L and Sattin M (2011) Characterisation and management of *Phalaris paradoxa* resistant to ACCase-inhibitors. Crop Protect. 30: 293–299.

Crawford MH, Rincon-Florez V, Balzer A, Dang YP, Carvalhais LC, Liu H and Schenk PM (2015) Changes in the soil quality attributes of continuous no-till farming systems following a strategic tillage. Soil Research 53: 263–273.

Dang YP, Moody PW, Bell MJ, Seymour NP, Dalal RC, Freebairn DM and Walker SR (2015) Strategic tillage in no-till farming systems in Australia's northern grains-growing regions: II. Implications for Agronomy, Soil and Environment Soil and Tillage Research 152: 1150123.

Délye C (2013) Unravelling the genetic bases of non-target-site-based resistance (NTSR) to herbicides: a major challenge for weed science in the forthcoming decade. Pest Manage. Sci. 69: 176–187.

Diggle AJ and Neve P (2001) The population dynamics and genetics of herbicide resistance—a modeling approach. pp. 61–99. *In:* Powles SB and Shaner DL (eds.). Herbicide Resistance and World Grains. Boca Raton: CRC Press Inc.

Dowling PM and Nicol HI (1993) Control of annual grasses by spraytopping and the effect on triticale grain yield. Aust. J. Agric. Res. 44: 1959–1969.

Duke SO (2012) Why have no new herbicide modes of action appeared in recent years? Pest Management Science 68: 505–512.

FAO (2015) Introduction to conservation agriculture (its principles & benefits). Available at http://teca.fao.org/technology/introduction-conservation-agriculture-its-principles-benefits. Accessed August, 2015.

Fleet B and Gill G (2012) Seed dormancy and seedling recruitment in smooth barley (*Hordeum murinum* ssp. glaucum) populations in southern Australia. Weed Sci. 60: 394–400.

Flower KC, Cordingley N, Ward PR and Weeks C (2012) Nitrogen, weed management and economics with cover crops in conservation agriculture in a Mediterranean climate. Field Crops Res. 132: 63–75.

Gill GS and Holmes JE (1997) Efficacy of cultural control methods for combating herbicide-resistant *Lolium rigidum*. Pestic. Sci. 51: 352–358.

Gommers CMM, Visser EJW, Onge KRS, Voesenek LACJ and Pierik R (2013) Shade tolerance: when growing tall is not an option. Trends Plant Sci. 18: 65–71.

Gould F (1991) The evolutionary potential of crop pests. Am. Sci. 79: 496–507.

Green JM (2014) Current state of herbicides in herbicide-resistant crops. Pest Manage. Sci. 70: 1351–1357.

Heap IM (2015) The International survey of herbicide resistant weeds. Available at http://www.weedscience.com. Accessed 31st August, 2015.

Hoyle JA, McElroy JS and Rose JJ (2012) Weed control using an enclosed thermal heating apparatus. Weed Technol. 26: 699–707.

Jana S and Thai KM (1987) Patterns of changes of dormant genotypes in *Avena fatua* populations under different agricultural conditions. Canadian Journal of Botany 65: 1741–1745.

Jasieniuk M, Brûlé-Babel AL and Morrison IN (1996) The evolution and genetics of herbicide resistance in weeds. Weed Sci. 44: 176–193.

Jordan NR and Jannink JL (1997) Assessing the practical importance of weed evolution: a research agenda. Weed Res. 37: 237–246.

Kassam A, Friedrich T, Derpsch R, Lahmar R, Mrabet R, Basch G, González-Sánchez EJ and Serraj R (2012) Conservation agriculture in the dry Mediterranean climate. Field Crops Res. 132: 7–17.

Kercher S and Conner JK (1996) Patterns of genetic variability within and among populations of wild radish, *Raphanus raphanistrum* (Brassicaceae). Am. J. Bot. 83: 1416–1421.

Kim DS, Brain P, Marshall EJP and Caseley JC (2002) Modelling herbicide dose and weed density effects on crop: weed competition. Weed Res. 42: 1–13.

Kleemann SGL and Gill GS (2006) Differences in the distribution and seed germination behaviour of populations of *Bromus rigidus* and *Bromus diandrus* in South Australia: adaptations to habitat and implications for weed management. Aust. J. Agric. Res. 57: 213–219.

Kloot P (1983) The genus *Lolium* in Australia. Aust. J. Bot. 31: 421–435.

Kobayashi H, Miura S and Oyanagi A (2004) Effect of winter barley as a cover crop on the weed vegetation in a no-tillage soybean. Weed Biology and Management 4: 195–205.

Kon KF and Blacklow WM (1989) Identification, distribution and population variability of great brome (*Bromus diandrus* Roth.) and rigid brome (*Bromus rigidus* Roth.). Aust. J. Agric. Res. 39: 1039–1050.

Lemerle D, Cousens RD, Gill GS, Peltzer SJ, Moerkerk M, Murphy CE, Collins D and Cullis BR (2004) Reliability of higher seeding rates of wheat for increased competitiveness with weeds in low rainfall environments. J. Agric. Sci. 142: 395–409.

Lemerle D, Gill GS, Murphy CE, Walker SR, Cousens RD, Mokhtari S, Peltzer SJ, Coleman R and Luckett DJ (2001) Genetic improvement and agronomy for enhanced wheat competitiveness with weeds. Aust. J. Agric. Res. 52: 527–548.

Leys AR, Cullis BR and Plater B (1991) Effect of spraytopping applications of paraquat and glyphosate on the nutritive value and regeneration of vulpia (*Vulpia bromoides* (L.) S.F. Gray). Australia Journal of Agricultural Research 42: 1405–1415.

Liebl RA and Worsham AD (1983) Tillage and mulch effects on morningglory (*Ipomoea* spp.) and certain other weed species. Proceedings of the Southern Weed Science Society 36: 405–414.

Liu H, Lee SH and Saunders C (2014) Development of a machine vision system for weed detection during both of off-season and in-season in broadacre no-tillage cropping lands. American Journal of Agricultural and Biological Sciences 9: 174–193.

Llewellyn RS, D'Emden FH and Kuehne G (2012) Extensive use of no-tillage in grain growing regions of Australia. Field Crops Res. 132: 204–212.

Londo JP and Schaal BA (2007) Origins and population genetics of weedy red rice in the USA. Mol. Ecol. 16: 4523–35.

Lutman PJW, Moss SR, Cook S and Welham SJ (2013) A review of the effects of crop agronomy on the management of *Alopecurus myosuroides*. Weed Res. 53: 299–313.

Marx C, Barcikowski S, Hustedt M, Haferkamp H and Rath T (2012) Design and application of a weed damage model for laser-based weed control. Biosys. Eng. 113: 148–157.

Mas MT and Verdú AMC (2003) Tillage system effects on weed communities in a 4-year crop rotation under Mediterranean dryland conditions. Soil Tillage Res. 74: 15–24.

Matthews JM, Llewellyn R, Powles S and Reeves T (1996) Integrated weed management for the control of herbicide resistant annual ryegrass. pp. 417–420. *In* Proceedings of the 8th Australian Agronomy Conference, Toowoomba, Queensland, Australia, 30 January-2 February (1996) Toowoomba, Australia: Australian Society of Agronomy.

McGowan A (1970) Comparative germination patterns of annual grasses in north-eastern Victoria. Aust. J. Exp. Agric. 10: 401–404.

McLachlan SM, Murphy SD, Tollenaar M, Weise SF and Swanton CJ (1995) Light limitation of reproduction and variation in the allometric relationship between reproductive and vegetative biomass in Amaranthus retroflexus (redroot pigweed). J. Appl. Ecol. 32: 157–165.

McLean AR, Widderick MJ and Walker SR (2012) Strategic tillage reduces emergence of key sub-tropical weeds. pp. 248. *In* Proceedings of the Proceedings of the 18th Australian Weeds Conference. Melbourne, Australia: Weed Society of Victoria.

Melander B, Lattanzi B and Pannacci E (2015) Intelligent versus non-intelligent mechanical intra-row weed control in transplanted onion and cabbage. Crop Protection 72: 1–8.

Menchari Y, Délye C and Le Corre V (2007) Genetic variation and population structure in black-grass (*Alopecurus myosuroides* Huds.), a successful, herbicide-resistant, annual grass weed of winter cereal fields. Mol. Ecol. 16: 3161–3172.

Mengistu LW and Messersmith CG (2002) Genetic diversity of kochia. Weed Sci. 50: 498–503.

Mohler CL and Teasdale JR (1993) Response of weed emergence to rate of *Vicia villosa* Roth and *Secale cereale* L. residue. Weed Research 33: 487–499.

Morgan PW, Finlayson SA, Childs KL, Mullet JE and Rooney WL (2002) Opportunities to improve adaptability and yield in grasses: Lessons from sorghum. Crop Sci. 42: 1791–1799.

Naylor JM and Jana S (1976) Genetic adaptation for seed dormancy in Avena fatua. Canadian Journal of Botany 54: 306–312.

Neve P and Powles S (2005) Recurrent selection with reduced herbicide rates results in the rapid evolution of herbicide resistance in Lolium rigidum. TAG Theoretical and Applied Genetics 110: 1154–1166.

Neve P, Vila-Aiub M and Roux F (2009) Evolutionary-thinking in agricultural weed management. New Phytol. 184: 783–793.

Norris RF (2007) Weed fecundity: Current status and future needs. Crop Protect. 26: 182–188.

Olfert MR, Stumborg M, Craig W and Schoney RA (1991) The economics of collecting chaff. Am. J. Alternative Agric. 6: 154–160.

Ozpinar S (2006) Effects of tillage systems on weed population and economics for winter wheat production under the Mediterranean dryland conditions. Soil Tillage Res. 87: 1–8.

Pearce GA and Holmes JE (1976) The control of annual ryegrass. Journal of Agriculture Western Australia 17: 77–82.

Pedersen BP, Neve P, Andreasen C and Powles SB (2007) Ecological fitness of a glyphosate-resistant Lolium rigidum population: Growth and seed production along a competition gradient. Basic Appl. Ecol. 8: 258–268.

Pérez-de-Luque A and Rubiales D (2009) Nanotechnology for parasitic plant control. Pest Manage. Sci. 65: 540–545.

Powles SB (1986) Appearance of a biotype of the weed, *Hordeum glaucum* Steud., resistant to the herbicide paraquat. Weed Res. 26: 167–172.

Powles SB and Yu Q (2010) Evolution in action: plants resistant to herbicides. Annu. Rev. Plant Biol. 61: 317–347.

Preston C and Powles SB (2002) Evolution of herbicide resistance in weeds: initial frequency of target site-based resistance in acetolactate synthase-inhibiting herbicides in *Lolium rigidum*. Heredity 88: 8–13.

Putnam AR and DeFrank J (1983) Use of phytotoxic plant residues for selective weed control. Crop Protection 2: 173–181.

Reeves TG and Smith IS (1975) Pasture management and cultural methods for the control of annual ryegrass (*Lolium rigidum*) in wheat. Aust. J. Exp. Agric. Anim. Husb. 15: 527–530.

Renton M and Flower KC (2015) Occasional mouldboard ploughing slows evolution of resistance and reduced long-term weed populations in no-till systems. Agricultural Systems 139: 66–75.

Roush ML and Radosevich SR (1985) Relationship between growth and competitiveness of four annual weeds. J. Appl. Ecol. 22: 895–905.

Scotford IM and Miller PCH (2005) Applications of spectral reflectance techniques in northern european cereal production: A review. Biosys. Eng. 90: 235–250.

Shirtliffe SJ and Entz MH (2005) Chaff collection reduces seed dispersal of wild oat (*Avena fatua*) by a combine harvester. Weed Sci. 53: 465–470.

Steadman KJ, Eaton DM, Plummer JA, Ferris DG and Powles SB (2006) Late-season non-selective herbicide application reduces *Lolium rigidum* seed numbers, seed viability, and seedling fitness. Aust. J. Agric. Res. 57: 133–141.

Teasdale JR (1993) Interaction of light, soil moisture, and temperature with weed suppression by hairy vetch residue. Weed Science 41: 46–51.

Van Acker RC (2009) Weed biology serves practical weed management. Weed Res. 49: 1–5.

Vandenbussche F, Pierik R, Millenaar FF, Voesenek LACJ and Van Der Straeten D (2005) Reaching out of the shade. Curr. Opin. Plant Biol. 8: 462–468.

Vigneault C, Benoit DL and McLaughlin NB (1990) Energy aspects of weed electrocution. pp. 12–26. Reviews of Weed Science.

Walsh MJ, Harrington RB and Powles SB (2012) Harrington seed destructor: A new nonchemical weed control tool for global grain crops. Crop Sci. 52: 1343–1347.

Walsh MJ, Maguire N and Powles SB (2009) Combined effects of wheat competition and 2.4-D amine on phenoxy herbicide resistant *Raphanus raphanistrum* populations. Weed Res. 49: 316–325.

Walsh MJ and Minkey DM (2006) Wild radish (*Raphanus raphanistrum* L.) development and seed production in response to time of emergence, crop topping and sowing rate of wheat. Plant Protection Quarterly 21(1): 25–29.

Walsh MJ, Newman P and Powles SB (2013) Targeting weed seeds in-crop: A new weed control paradigm for global agriculture. Weed Technol. 27: 431–436.

Walsh MJ and Powles SB (2007) Management strategies for herbicide-resistant weed populations in Australian dryland crop production systems. Weed Technol. 21: 332–338.

Walsh MJ and Powles SB (2009) Impact of crop-topping and swathing on the viable seed production of wild radish (*Raphanus raphanistrum*). Crop Pasture Sci. 60: 667–674.

Webster TM and Grey TL (2015) Glyphosate-resistant palmer Amaranth (Amaranthus palmeri) morphology, growth, and seed production in Georgia. Weed Sci. 63: 264–272.

Wu LIN, Till-Bottraud I and Torres A (1987) Genetic differentiation in temperature-enforced seed dormancy among golf course populations of Poa annua L. New Phytol. 107: 623–631.

Yenish JP and Young FL (2004) Winter wheat competition against jointed goatgrass (*Aegilops cylindrica*) as influenced by wheat plant height, seeding rate, and seed size. Weed Sci. 52: 996–1001.

Zerner MC, Gill GS and Vandeleur RK (2008) Effect of height on the competitive ability of wheat with oats. Agron. J. 100: 1729–1734.

Zimdahl RL (2007) Weed-Crop Competition: A Review. Hoboken: Wiley.

Zimdahl RL (2013) Fundamentals of Weed Science California, USA: Academic Press.

Chapter 5

Evolution of Weed Resistance to Herbicides

What Have We Learned After 70 Years?

Stephen O. Duke[1,*] and *Ian Heap*[2]

Introduction

Early Developments

Sufficiently strong and constant selection pressure on a population of organisms over an adequate time period inevitably leads to compensatory genetic changes in that population. Charles Darwin understood this in general terms more than 150 years ago. The larger and the more fecund the population exposed to the selection pressure, the more rapidly this evolutionary process occurs for a species. Also, the frequency and duration of exposure to the selection pressure is positively correlated with the speed of evolved resistance. The advent of modern, synthetic pesticides fulfilled all of these requirements for rapid evolution of pesticide resistance in pests. Just as with antibiotics, any evolutionary biologist could have foretold the result of extensive and constant use of pesticides—widespread and relatively rapid evolution of resistance.

---

[1] Research Leader, Natural Product Utilization, Agricultural Research Service, United States Department of Agriculture, Cochran Research Center, P. O. Box 1848, University, MS 38677, USA.
[2] Director, International Survey of Herbicide-Resistant Weeds, P. O. Box 1365, Corvallis, OR 97339, USA.
Email: IanHeap@weedscience.org
* Corresponding author: Stephen.Duke@ars.usda.gov

After more than 10,000 years of agriculture with labor-intensive weed management, the introduction of synthetic herbicides in the mid-1940s offered farmers a cost-effective and simple method of weed management with greatly reduced manual labor for the first time. So, this technology was rapidly adopted in countries that could afford it, beginning with auxinic herbicides such as 2,4-D, followed by hundreds of compounds with more than twenty modes of action (MOAs) being commercialized over the past 70 years (Fig. 1).

The first cases of evolved herbicide resistance were to one of the first commercialized synthetic herbicides, 2,4-D (Timmons 2005). Because few synthetic herbicides were available during the early days of 2,4-D use, it was widely used for control of broadleaf weeds. Wild carrot (*Daucus carota* L.) (Switzer 1957) and spreading dayflower (*Commelina diffusa* Burm. f) (Hilton 1957) were both reported to have evolved resistance to 2,4-D in 1957, around 10 years after use of 2,4-D had become widespread. Evolved resistance in these weed species was not a major problem as herbicides with new MOAs were being commercialized at a rapid pace (Fig. 1). Therefore, these reports did not cause much concern.

The next herbicide MOA group to which weeds evolved resistance was that of photosystem II (PSII) inhibitors. These herbicides were not introduced until the early 1950s (Fig. 1), and the first case of evolved

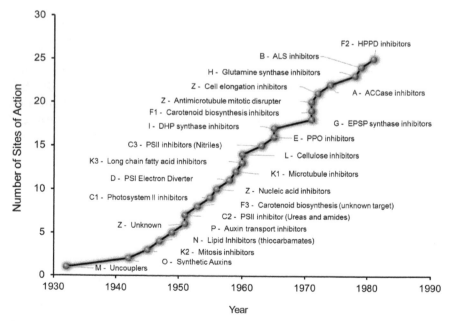

Figure 1. Introduction of new herbicide modes of action over the years. Letters are HRAC herbicide class designators.

herbicide resistance was not confirmed until 1970 when triazine-resistant common groundsel (*Senecio vulgaris* L.) was reported by Ryan (1970). Thereafter, other photosystem II (PSII) inhibitor-resistant weed species began to increase rapidly, including resistances to other chemical classes of PSII herbicides (Fig. 2). At this time (the early 1970s), more than half of the 25 or so herbicide MOAs had been introduced (Fig. 1), so those involved in weed management still had many options to replace or augment herbicides that were less useful due to evolved resistance.

The Linear Phase of New Herbicide Resistance Cases

At this time (the mid-1970s), a linear increase of evolved herbicide resistance began that has continued unabated until the present time, now approaching almost 500 unique cases of evolved weed resistance (Fig. 3). This linear increase includes data from the rapid evolution of resistance to some herbicide classes, such as the acetolactate synthase (ALS) and acetyl CoA carboxylase (ACCase) inhibitors, as well as data from the slower evolution of resistance to other classes of herbicides such as auxinic herbicides (Fig. 2). Beckie (2006) categorized the risk of evolution of herbicide resistance by herbicide MOA (Table 1). Based on newer global resistance information (Heap 2015), 4-hydroxylphenylpyruvate dioxygenase (HPPD)

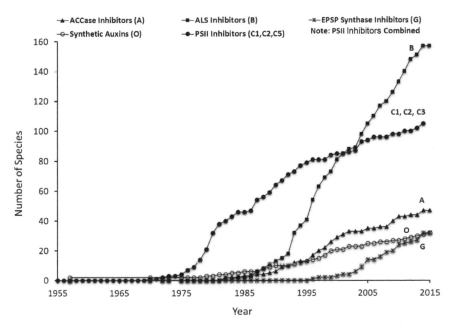

Figure 2. Chronological evolution of herbicide resistance to five herbicide modes of action.

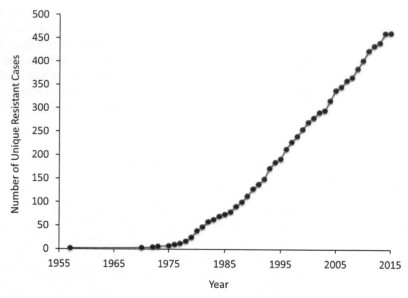

Figure 3. Chronological, global increase in number of unique cases (species and mode of action) of herbicide resistance.

Table 1. Risk for evolution of resistance by mode of action and HRAC (2015) classification according to Beckie (2006).

High

A – ACCase inhibitors (e.g., cyclohexanediones)

B – ALS inhibitors (e.g., sulfonyureas)

Moderate to high

C1 – PSII inhibitors (e.g., triazines)

Moderate

K1 – microtubule assembly inhibitors (e.g., dinitroanilines)

N – Non-ACCase fatty acid synthesis (e.g., thiocarbamates)

Low to moderate

C2 – PSII inhibitors (e.g., ureas)

D – PSI electron diverters (bipyridyliums)

Low

O – Auxinic herbicides (e.g., 2,4-D)

C3 – PSII inhibitors (e.g., nitriles)

G – EPSPS inhibitor (glyphosate)

H – GS inhibitor (glufosinate)

and protoporphyrinogen oxidase (PPO) inhibitors can be added to the relatively low risk herbicide classes. There are other classes, such as the phytoene desaturase (PDS) inhibitors and the one serine/threonine protein phosphatase inhibitor (Bajsa et al. 2012), endothall, for which there are very few species with evolved resistance. In the case of PDS inhibitors, this may be due to the requirement for two base pair mutations in a single codon to get an amino acid change that confers resistance in most plant species (Dayan et al. 2014). Endothall apparently inhibits all of the several (more than 20 in Arabidopsis) serine/threonine protein phosphatases in plants (Basja et al. 2012), making target site resistance in one of these enzymes unlikely to provide a significant level of resistance.

Different herbicide classes have been unequally used, so that selection pressure from the different classes has been very uneven. This makes it difficult to judge the risk of herbicide evolution for the little-used herbicide classes for which there has been little or no reported resistance. Also, the effective duration of herbicidal doses of different herbicides varies considerably, with that of some soil-applied herbicides lasting several weeks, while the effective duration of a dose of some foliar-applied herbicides lasts a shorter time. Furthermore, plants that do not intercept foliar herbicides generally escape exposure to effective doses of the herbicide.

An attempt was made to compare the frequency of resistance mutations to two ALS-inhibiting herbicides and glyphosate in chemically-mutagenized *Arabidopsis* (Jander et al. 2003). In 250,000 mutagenized plants, no glyphosate-resistant mutants were found, but four plants resistant to ALS inhibitors were found, leading the investigators to conclude that evolution of resistance to ALS inhibitors occurs more easily than to glyphosate. This conclusion is strongly supported by what has happened in the field (Heap 2015). For example, ALS inhibitor herbicides were introduced 8 years (1982) after the introduction of glyphosate (1974) (Duke et al. 2003; Tranel and Wright 2002), yet weeds resistant to ALS inhibitors appeared first in 1987 (Mallory-Smith et al. 1990), followed by the first report of a glyphosate-resistant weed in 1996 (Pratley et al. 1999). Thus, the lag time between introduction and first resistance was 5 and 22 years, respectively, for ALS inhibitors and glyphosate. Because both types of herbicides were used extensively after introduction, the more than four-fold difference in lag time for resistance to appear is best explained by the fact that any one of several base pair mutations will confer target site resistance to ALS inhibitors (Tranel and Wright 2002; Yu and Powles 2014a), whereas, two simultaneous base pair mutations in different codons are needed for a high level of resistance at the target molecule (5-enolpyruvylshikimate-3-phosphate synthase; EPSPS) for classical target site resistance to glyphosate (Bradshaw et al. 1997). However, a single base pair mutation in EPSPS provides weak resistance to glyphosate for populations of several weed species (Sammons and Gaines 2014), a result that was not anticipated by Bradshaw et al. (1997).

For the first 40 years of herbicide use, a new MOA appeared every 2 to 3 years, giving farmers more options for weed control when resistance evolved (Fig. 1). Unfortunately, introduction of new modes action came to an abrupt halt around 1985 (Duke 2012; Gerwick 2010), when the linear increase in evolved herbicide resistance had been occurring for less than 10 years (Fig. 3). The HPPD inhibitors were the last clearly established MOA. So, as the need for new MOAs increased rapidly due to evolved resistance, no more were offered. This scenario led to an increasing need for more diversified and complex (expensive) weed management strategies in the early 1990s, which made the prospect of a simple and effective technology for killing all weed species with only one herbicide highly attractive. Glyphosate used with glyphosate-resistant (GR) crops fulfilled this need.

Impact of Glyphosate and Glyphosate-Resistant Crops

About a decade after the last new herbicide MOA was introduced, GR crops were introduced in 1996 (Duke 2014; Duke and Powles 2008). Glyphosate is a non-selective herbicide, so it kills almost all weed species at recommended application rates. It is efficient at killing perennial weeds because it is slow acting and highly phloem mobile with preferential translocation to metabolic sinks, such as apical meristems (Duke et al. 2003). Thus, it prevents regrowth from meristems that are physically remote from parts of the plant treated with the herbicide. Glyphosate has been considered the most perfect herbicide ever discovered for these and other reasons (Duke and Powles 2008). The only significant problem with glyphosate is that it is practically inactive in soil, due to its tight binding to soil components (Duke 1988). Thus, it can only be used as a foliar spray, so weeds that emerge after spraying are not controlled. Nevertheless, the effectiveness, simplicity, and relatively low cost for weed management with glyphosate in GR crops resulted in their rapid and almost universal adoption where they were available. After the glyphosate patent expired in 2000, glyphosate became a less expensive, generic herbicide, making use of GR crops even more economically attractive.

The phenomenal success of this crop/herbicide combination reduced the use and the value of the global market for other herbicides in maize (*Zea mays* L.), soybean (*Glycine max* (L.) Merr.), cotton (*Gossypium hirsutum* L.), canola (*Brassica napus* L.), sugarbeets (*Beta vulgaris* L.), and alfalfa (*Medicago sativa* L.) (Duke 2014; Shaner 2000). Extensive adoption of glyphosate in GR crops and the concomitant drastic reduction of the use of other herbicides with these crops led to two events that strongly influenced evolution of herbicide resistance.

The first was a reduction in research to discover new herbicides, thus reducing the chances that a viable, new herbicide MOA would be discovered or introduced if discovered. Both the number of companies conducting herbicide discovery research and the amount of herbicide discovery research in surviving companies were reduced (Duke 2012). These reductions were both influenced by the dominance of glyphosate. The lowered value of the non-glyphosate herbicide market made the introduction of a new herbicide, even with a new MOA, less likely. In fact, whether some of the herbicides introduced before GR crops would have been commercialized after 1996 is questionable. So, some herbicides that might be useful now may have been passed over during the time between 1996 and the recent rise of GR weeds, a time when the patent right period of such compounds could have expired or dwindled to such a short time that commercialization would be too economically risky.

The second event was the use of only one herbicide, glyphosate, for weed control in GR crops by many farmers for the first decade or so of GR crop availability. This "perfect storm" for evolution of resistance (strong selection pressure of a single lethal agent with a single MOA for several years on highly fecund pests) led to the evolution of resistance to a herbicide for which the evolution of resistance was predicted to occur slowly and at a low level for good genetic and physiological reasons (Bradshaw et al. 1997). As one might have predicted, much of the rapid rise in evolution of glyphosate resistance (Fig. 4) has been in GR crops in the U.S., Argentina, and Brazil (Heap 2014), where GR crops have been massively adopted (James 2014). Even before GR crops, glyphosate was used extensively in plantation crops such as fruit trees, where application to herbaceous weeds can be done several times a year without significant damage to the crop. This type of repeated use pattern resulted in some of the earlier GR weed cases. For example, GR *Eleusine indica* L. evolved in a fruit orchard (Lee and Ngim 2000), and GR *Lolium rigidum* Gaud. also evolved in an orchard (Powles et al. 1998). Glyphosate was seldom used more than two times a year in GR crops, but the selection pressure has usually been uninterrupted for years, and the vast areas planted with GR crops led to selection pressure on vast numbers of weeds of many species. The result was evolution of a growing number of GR weed species in GR cropping situations. This occurred despite farmers being advised to use best management practices such as alternating herbicide MOAs, using more than one MOA for all weed species in a field in a single year, and/or using supplemental tillage or cultural methods (Frisvold et al. 2009; Norsworthy et al. 2012; Powles et al. 1997). Overuse of this "once in a century" herbicide (Duke and Powles 2008) has reduced the value of GR crops and glyphosate, the most perfect weed management tool yet devised.

The question as to whether GR crops affected evolution of resistance to herbicides with other MOAs is often asked. The global increase in unique

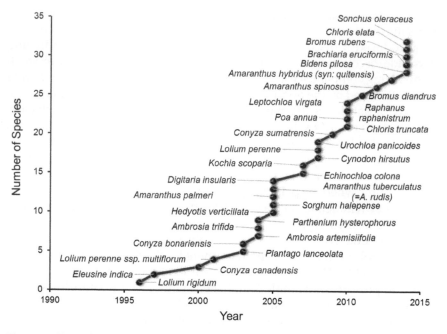

Figure 4. Chronological worldwide accumulation of weed species with evolved resistance to glyphosate.

herbicide resistance cases continued unabated after 1996, when GR crops were introduced, suggesting that GR crops have had little impact on this generally linear increase in herbicide resistance that began around 1980 (Fig. 3). Perhaps without GR crops the increase would have accelerated due to greater use of herbicides with MOAs to which resistance evolves more rapidly than to glyphosate. But, the data suggest that, at least on a global scale, the overall impact of GR crops on evolution of resistance to herbicides other than glyphosate has been small.

Mechanisms of Resistance

We will only make some general comments about mechanisms of resistance, as other chapters in this book will provide much more detail. Resistance mechanisms can be target site- or non-target site-based. In the case of target site resistance, there is no alteration in movement to the target site, but after arriving at the target site, there is less effect, either due to an alteration in the target enzyme affinity for the herbicide or a greater number of target sites (e.g., gene amplification), thus diluting the herbicide. Target site alteration is most commonly due to a single base pair alteration in a codon that changes one amino acid in the target protein, as has been commonly found with resistance to PSII, ACCase, and ALS inhibitors. There are several base pair combinations for some amino acids (http://www.cbs.dtu.dk/courses/27619/codon.html). Thus, in some cases, two base pair changes in the same codon are needed to provide the codon needed for the amino acid that provides resistance (e.g., Dayan et al. 2014). Such a double mutation in a single codon is exponentially more improbable than a one codon mutation. But, highly improbable events can occur. For example, an entire codon deletion occurred to provide target site resistance to PPO inhibitors (Patzoldt et al. 2006). Another unlikely event is for a change in two base pairs to alter two amino acids in ways that provide the changes in EPSPS needed for a high level of resistance to glyphosate (Yu et al. 2015). The extremely small chance for this was the reason given decades ago for the improbable evolution of significant resistance to glyphosate (Bradshaw et al. 1997). These authors did not consider that "creeping" evolution of resistance can occur at the target site as well as with other mechanisms of resistance. Thus, a single codon alteration can provide enough resistance for sufficient survivors to field rates of the herbicide, and these survivors can be selected for higher levels of resistance via another base pair change when higher field rates are used to control the populations with a low resistance level. Very low probability events can occur with strong and prolonged selection pressure over vast areas.

Gene amplification that results in enough functional gene copy numbers for a herbicide target site was another unexpected mechanism of herbicide resistance (Gaines et al. 2010). Perhaps this should not have been a surprise, as this mechanism of insecticide resistance has been known for some time for both target site and non-target site resistance to insecticides (Bass and Field 2011; Devonshire and Field 1991).

Non-target site resistance mechanisms reduce the amount of herbicide getting to the target site. These include reduced uptake and/or translocation, sequestration into parts of the plant with little or no target site, and metabolic conversion of the herbicide molecule to an inactive molecule. Sequential, "creeping" evolution of herbicide resistance is more common with this mechanism, as it usually involves multiple genes functioning in metabolism and/or sequestration of the herbicide. Thus, reduced rates of a herbicide in the field should favor this mechanism of resistance, as weed survivors that have evolved only a low level of resistance can be further selected for higher levels of resistance by mutations in other genes that can contribute to resistance (Gardner et al. 1998; Yu et al. 2013). Low doses may also increase mutation rates in weeds that are stressed by sublethal exposure to the herbicide (Gressel 2011).

However, in the same fields in which full rates are applied, one species can evolve non-target resistance, and another species can evolve target site resistance to the same herbicide. For example, the mechanism of resistance to glyphosate in horseweed (*Conyza canadensis* L. Cronq.) is sequestration of glyphosate into the vacuole (Ge et al. 2010), whereas in the same geographical area where this mechanism evolved (the US Southeast), EPSPS gene amplification has evolved to provide resistance to Palmer amaranth (*Amaranthus palmeri* S. Wats.) (Gaines et al. 2010). So, there can be exceptions to generalizations. Some populations of tall waterhemp (*Amaranthus tuberculatus* Moq. Sauer) have two resistance mechanisms, an EPSPS gene mutation that provides a low level of resistance and reduced uptake and translocation of glyphosate (Nandula et al. 2013). We do not know which came first, but this would be another case of 'creeping resistance', where one mutation would provide enough resistance for survivors to be selected for more robust resistance by adding another mechanism of resistance.

Charles Darwin would have been impressed with the varied genetic responses of weeds to herbicides. Weeds have been much more resilient to the selection pressure imposed by herbicides than the pioneers of herbicide-based weed management would have envisioned.

Multiple and Cross Resistances

As new cases of herbicide resistance accumulate at a rate of about 12 per year (Fig. 3), some weed species have accumulated resistance to more than one herbicide MOA, each by a different mechanism of resistance (multiple herbicide resistance), in some cases with resistance to several MOAs (Fig. 5). The different resistance mechanisms in multiple resistant weed populations evolve separately, either with evolution of resistance to a new MOA in a population that already has resistance to another MOA or through gene flow between populations where each population has independently evolved resistance to one MOA. With multiple resistance, resistance to more than one MOA can be found in individual plants in a population. These cases can involve just target sites mechanisms (Bi et al. 2015), just non-target-site mechanisms (Huffman et al. 2015), or both target site and non-target site mechanisms (e.g., Bell et al. 2013). Considering all of the resistance mechanisms to the many commercial herbicides that exist, the multiple resistance problem is growing at a rapid pace and will probably become much worse, leaving farmers with little or no herbicide options.

A single mechanism of resistance can often provide resistance to other herbicides with the same MOA and sometimes to herbicides with different MOAs (cross resistance). Cross resistance to herbicides of the same class is common with target site resistance. For example, one target protein gene change can result in resistance to some, but not all, herbicides with the same MOA for PSII (Fuerst et al. 1986), PDS (Arias et al. 2006), ACCase (Kaundun 2014), and ALS (Yu and Powles 2014a) inhibitors, to give just few examples. In such cases, negative cross resistance, the enhanced sensitivity

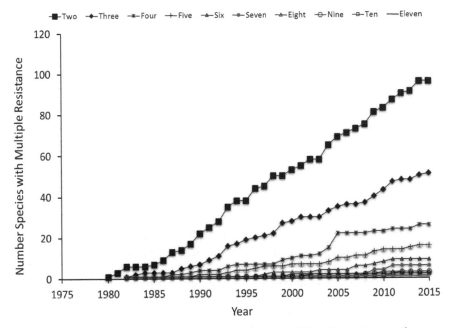

Figure 5. Chronological increases over time of weed species with resistance to more than one site of action.

to a herbicide from the herbicide class to which resistance has evolved, is sometimes found. For example, Arias et al. (2006) found a mutation in a *Hydrilla verticillata* L. f. Royle PDS gene (Arg304 conversion to Thr) that gave a 52-fold resistance of the enzyme to norflurazon, but rendered the enzyme 5-fold more sensitive to diflufenican. Similarly, Fuerst et al. (1986) reported a mutation in common groundsel (*Senecio vulgaris* L.) that produced plants that were 1,500-fold more resistant to atrazine than the wild type, but were about eight-fold more sensitive to dinoseb. There were other examples of negative cross resistance in both of these papers. Cross resistance is also common with non-target site resistance, especially with metabolic degradation mechanisms, as many of the enzymes involved in this type of mechanism (e.g., glutathione-S-transferases, P450 monooxygenases, glycosyl transferases) have many possible substrates, so that they can transform xenobiotic molecules from more than one chemical family. For example, *Lolium rigidum* Gaud. and *Alopecurus myosuroides* Huds. have evolved cross resistance to different herbicide classes with different MOAs by a common mechanism of resistance—enhanced metabolic degradation (Yu and Powles 2014b). In the case of *L. rigidum*, evolved, enhanced metabolic degradation of diclofop apparently involves enhanced gene expression of at least three enzyme types, cytochrome P450s, a glutathione transferase, and a nitronate monooxygenase (Gaines et al. 2014).

What We Do Not Know

We are still woefully ignorant of many aspects of herbicide resistance. Why are some plant species predisposed to evolve resistance to certain herbicides or to many herbicides? For example, three *Conyza* species have each evolved resistance to the same four MOAs, with some individual populations having resistance to at least two of these MOAs (Heap 2015). The situation with *L. rigidum* is even worse, with some populations having multiple resistance to seven sites of action, although some of this is actually metabolic-based cross resistance (Yu and Powles 2014b). *Amaranthus* species have accumulated resistance to a large array of herbicides over the years with many cases of multiple resistance (Heap 2015). In the same fields, other species have evolved little or no resistance to herbicides.

Why are certain mechanisms of resistance favored over others for certain herbicides? Virtually all selective herbicides owe their crop selectivity to rapid metabolic degradation of the herbicide. This would suggest that weeds would preferentially evolve similar mechanisms of resistance. This does not appear to be the case at first glance. Although non-target site mechanisms of resistance are common, most of the clearly documented herbicide resistance mechanisms reported are target site-based. This is probably partly because scientists prefer to study and document cases of high levels of resistance which are mostly due to target site alteration. Low levels of resistance are inherently more difficult to study. Furthermore, multigene-based metabolic degradation begins at a low level until selection results in populations with enough genes in place for robust resistance. This phenomenon masks many populations of weeds with low resistance levels based on herbicide metabolism or other non-target site mechanisms.

That glyphosate resistance based on metabolism has not been clearly documented is surprising, in that we know that many plants can metabolically degrade the herbicide into aminomethylphosphonic acid (AMPA) and glyoxylate (Duke 2010; Reddy et al. 2008), and evolution of target site resistance and other non-metabolic means of resistance has not been simple (Bradshaw et al. 1997; Gaines et al. 2010; Ge et al. 2010; Yu et al. 2015). Enhanced metabolic degradation of a herbicide like glyphosate that acts much more slowly than most other herbicides would seem to be favored as the mechanism of resistance, as even a slightly enhanced rate of degradation might effectively reduce the herbicide concentration to sub-lethal levels. Yet, there is not one clearly confirmed case in which a weed species has evolved resistance to glyphosate by enhanced metabolic degradation. Perhaps the weakly phytotoxic properties of AMPA (Reddy et al. 2004) have reduced the probability of such a mechanism. No GR crops have been produced with only a microbial transgene for glyphosate oxidoreductase (GOX) that produces AMPA and glyoxylate, even though such a transgene has been coupled with the *cp4* gene for a resistant EPSPS in GR canola (Duke 2014). This suggests that there is a problem with this mechanism of resistance alone.

For resistance mechanisms that involve site of action, we know why some target sites are favored for evolution of resistance, such as ACCase and ALS. To be succinct, the herbicide binding sites for these herbicides are highly promiscuous, with many effective herbicides being available for each MOA, and there are multiple one base pair changes that can provide resistance for each MOA without a significant alteration in the fitness of the mutant (Kaundun 2014; Tranel and Wright 2002; Yu and Powles 2014a). For ALS inhibitors, 157 species have evolved resistance, even including weed species with relatively low seed production, such as common cocklebur (*Xanthium strumarium* L.) (Heap 2015). For ACCase inhibitor herbicides, there are only 47 resistant plant species (Heap 2015), and some of the reported site of action alterations do confer a fitness penalty, such as in some mutants of *A. myosuroides* (Délye et al. 2013). This may explain why there are more cases of metabolic resistance to ACCase inhibitors (Ahmad-Hamdani et al. 2012; Délye et al. 2007) than to ALS inhibitors. However, some metabolic resistance is a secondary effect of selection with other herbicides for which target site resistance does not evolve so easily.

Where Do We Go from Here?

A resistance crisis has evolved for all biocidal products (antibiotics and pesticides) that have become vital to the preservation of human health and well-being over the past century. The situation with herbicides may not be as dire as that with antibiotics, partly because other technologies are being developed that can replace or complement herbicides, whereas the medical options for antibiotics are not so obvious. But, in both cases, there is also renewed effort to find new biocides.

New transgenic herbicide-resistant crops made resistant to old herbicides are becoming available in the US (Duke 2014; Green 2014). Resistance to auxinic herbicides (2,4-D and dicamba), as well as to HPPD and ALS inhibitor herbicides, will be stacked with glyphosate and glufosinate resistance (USDA 2015). As mentioned earlier, the first case of herbicide resistance was to 2,4-D, and several other cases have evolved during the 70 years of 2,4-D use (Heap 2015). Dicamba resistance is already present. So, we can expect more cases of resistance to these herbicides after these herbicide-resistant crops become available. These new tools will have utility for many weed management situations, but we view them as short term solutions. There will probably be no large increase in herbicide-resistant crops, for many of the reasons discussed by Devine (2005) a decade ago. Limited consumer acceptance, cost of regulatory approval, and potential for economic reward does not provide sufficient impetus for most potential herbicide-resistant crops. Indeed, perusal of the USDA/APHIS website listing petitions for deregulation of herbicide-resistant crops clearly shows that relatively few herbicide-resistant crops are near introduction to the market. Non-transgenic methods of making crops resistant to herbicides, such as the imidazolinone-resistant crops (Duke 2014; Tan et al. 2005), have greater consumer acceptance and less cost for introduction. Modern precision gene editing and genome engineering methods such as CRISPR/Cas9, zinc finger nucleases (ZRN), transcription activator-like effector nucleases (TALEN), and oligonucleotide-directed mutagenesis (Gaj et al. 2013; Sauer 2015) offer tremendous opportunities for making crops resistant to herbicides (Voytas and Gao 2014). For example, CRISPR/Cas9 methods easily produced chlorsulfuron-resistant maize with no transgene by specific alternations in the ALS gene of choice (Svitashev et al. 2015) (Fig. 6). These technologies have the potential to economically provide crops resistant to any herbicide or combination of herbicides. However, current enthusiasm about these methods for crop improvement could be mitigated if regulatory

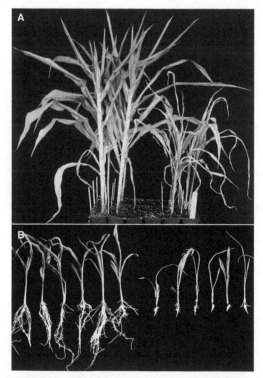

Figure 6. Maize plants with an ALS gene edited with CRISPR/Cas9 methods to provide resistance to chlorfulfuron on the left and wild type plants on the right. (A) Three weeks after four-week-old plants were sprayed with chlorsulfuron. (B) Fourteen-day old plants from embryos germinated on chlorsulfuron. With permission from Svitashev et al. (2015).

approval of such crops is complicated and/or public acceptance is no better than for transgenic crops.

The herbicide resistance crisis has caused renewed interest and effort from the pesticide industry in the discovery of herbicides with new MOAs. But, there are no clear new MOAs in the herbicides commercialized over the past ten years (Jeschke 2016). There are fewer herbicide than insecticide and fungicide MOAs, and new fungicide and insecticide MOAs have been introduced much more recently than the last new herbicide MOA (Duke 2015). Have we run out of new herbicide MOAs? Analysis of the MOAs of natural compounds that kill plants reveals that there are herbicide target sites that we have not yet exploited (Dayan and Duke 2014). If the primary role in nature of a compound that is phytotoxic is as a phytotoxin, it has evolved this function over many more years than the short time that synthetic herbicides have been used. If natural phytotoxins have retained their functions as phytotoxins over such long periods, evolution of resistance has not been as rapid or pronounced as we have seen with synthetic herbicides. One reason for this is that most of those natural phytotoxins that are still active may have more than one MOA, targeting multiple enzymes. This would at least make target site resistance highly unlikely. All of the MOAs of synthetic herbicides are considered to be associated with single target (HRAC 2015), although in some cases the single target is actually multiple proteins with similar functions (e.g., auxinic herbicides and endothall). Amitrole may actually have multiple MOAs (Devine et al. 1993). Despite 60 years of use, only four weed species have been reported to be resistant to amitrole (Heap 2015), but the mechanism(s) of resistance has not been reported.

Some of the natural phytotoxins discussed by Dayan and Duke (2014) are reported to have at least two molecular targets. For example, the plant pathogen-produced phytotoxin cyperin inhibits both PPO (Harrington et al. 1995) and enoyl (ACP) reductase (Dayan et al. 2008). Whether both MOAs are active at the same concentration of cyperin is not known. Another example is the *Sorghum*-produced allelochemical sorgoleone. It inhibits photosystem II, mitochondrial electron transport, HPPD, and root H^+-ATPase (Dayan et al. 2010). Dayan et al. (2009) concluded that PS II inhibition may be more important in young seedlings, and the other target sites may be more involved in the MOA in older plants. For such a molecule, a non-target site mechanism of resistance would be favored. For allelochemicals involved in plant-plant interactions, we are unaware of any resistance mechanisms that have been discovered, although implicit in the "novel weapons hypothesis" (Callaway and Ridenour 2014) is the assumption that plants in their native habitat have evolved some level of resistance to potential allelochemicals produced by their neighbors. Unfortunately, there is no rigorous evidence to validate the novel weapons hypothesis, especially proof that plants in the same native habitat have evolved resistance to each other's allelochemicals.

If most or many plant allelochemicals have multiple targets, metabolic degradation should be the mechanism of resistance most likely to evolve. We know that several plant species can metabolically inactivate benzoxazinoid allelochemicals (Baerson et al. 2005; Schulz et al. 2013). Nevertheless, benzoxazinoids are effective in suppressing many types of vegetation (Belz 2007). Target site resistance to natural phytotoxins used as virulence factors by plant pathogens is also rare, but resistance based on metabolism has been described. For example, wheat lines that are resistant to deoxynivalenol metabolize it to non-phytotoxic compounds (Lemmens et al. 2005). Nevertheless, what little we know of natural phytotoxins, suggests that they may often be more buffered from evolved resistance by means of multiple MOAs. Thus, new herbicides based on such compounds may have resistance management benefits that we have not experienced with existing synthetic herbicides.

The synthetic pesticide market is expected to grow for at least the next decade, but Olson (2015) has estimated that by the middle of the 21st century biopesticides may overtake synthetic pesticides in value. Currently, bioherbicides comprise a very small proportion of the biopesticide market. Biochemical bioherbicides have the most to offer with regard to fighting herbicide resistance. The US Environmental Protection Agency defines biochemical biopesticides as natural (biosynthesized) compounds that can be used for pest management. The momentum for discovery and introduction of biochemical biopesticides (Dayan and Duke 2014; Duke and Dayan 2015; Duke et al. 2014) may hasten the discovery of natural product-based herbicides with more than one MOA, or at least herbicides with new MOAs. But, without proper stewardship, non-target site resistance to herbicides with more than one mode action will occur.

Herbicide synergists that act by reducing herbicide metabolism in weeds, by increasing herbicide sequestration and/or protection from oxidative stress, or by altering herbicide stress signal transduction are a highly attractive technology for reducing herbicide rates (Shaner and Beckie 2014). But, in weed species in which the synergist does not work, the low rates would favor evolution of metabolic resistance, and perhaps increase evolution of resistance by increasing mutation frequencies (Gressel 2011).

It is unclear as to whether negative cross resistance has been explored by the agrochemical industry as a way of fighting resistance. There are few published comprehensive studies to determine the level of hypersensitivity to analogs of different chemistries with the same target site of herbicides to which resistance has evolved. In some cases of negative cross resistance, the resistant biotype is hypersensitive to commercially available herbicides. For example, some mutants of *Hydrilla verticillata* L. f. Royle with target site resistance to the PDS inhibitor fluridone are hypersensitive to the commercial PDS inhibitors beflubutamid, picolinafen and diflufenican (Arias et al. 2006). In other words, the mutation that reduces binding of fluridone increases binding of these other herbicides to PDS. So, having a PDS that is resistant to fluridone and all the other PDS-inhibiting herbicides may be virtually impossible. Rotation or mixing such herbicides should prevent evolution of target site resistance. The herbicides to which fluridone has negative cross resistance could be used at a reduced rate, as any plant with evolved target site resistance to fluridone would be hypersensitive. This is one case in which using herbicides with the same MOA sequentially might be justified.

Precision agriculture methods that deliver post emergence herbicides more accurately to weeds has the potential to greatly reduce the amount of herbicide used in a field. Real time detection of the exact location of specific weed species by incorporating image and location analysis into the spray machinery, so that the appropriate herbicide is accurately sprayed only on the proper weed species at the right dose for the growth stage is within the realm of possibility in the near future. Such technology would greatly reduce herbicide use, but the selection pressure as it relates to resistance management would still be an issue. The reward for investment in discovery and development efforts by the agrochemical industry would be reduced by this technology.

Precision, robotic weeding has been developed for organic crops, for which the cost of hand weeding is prohibitive (Fennimore et al. 2014). As the cost of this technology is reduced, it may supplant or augment herbicide use in non-organic horticultural crops and eventually agronomic crops. Weeds will evolve survival mechanisms for this technology also.

Sprayable RNAi has been proposed as a next generation method of weed management. The concept is to spray weeds with RNAi that targets the mRNA from a herbicide binding site gene such as that for EPSPS. The plant is thus deprived of sufficient enzymatic activity for normal functioning, resulting in stunted growth or death. Thus, in many ways it is like killing the plant with a herbicide that binds the same target. The RNAi can be used either with or without the complementary herbicide, but much less of the herbicide should be needed, in that the amount of target site is lowered by the RNAi. When targeting parasitic weeds such as *Orobanche aegyptiaca* Pers., RNAi that selectively blocks enzymatic function in the weed can be produced by the crop via a transgene (Aly et al. 2009). But, to use RNAi as a sprayed product there are technical problems in formulation of RNAi to efficiently get it into the target plant. Significantly more economical production of RNAi must also be developed. Unfortunately, there are no peer-reviewed research articles on the topic of sprayable RNAi for weed management, so we can only be assured by the fact that one major company is providing evidence in oral presentations that a major effort is being put into development of this technology.

Technically, any gene product of the plant can be targeted by RNAi, but one of the reasons many enzymes are not good herbicide targets is that there are too many copies of the enzyme to inhibit a sufficient percentage of them to kill the plant. For example, RUBP carboxylase is an essential enzyme for green plants, but it is present in huge amounts which would necessitate unreasonable amounts of a herbicide to poison enough of it to kill the plant. An analogous situation is the large amount of EPSPS found in weeds that have evolved resistance to glyphosate by gene amplification (Gaines et al. 2010; Jugulam et al. 2014; Salas et al. 2012). This will probably be an issue for some genes with RNAi also, although a company patented several new MOAs several years ago, based on gene silencing technology results (summarized by Duke et al. 2009). Those genes might be good targets for RNAi in weed management.

Again, if RNAi technology for weed management is successful, it may have a relatively short period of success if overused. Like herbicide target sites, mutations to change mRNA sequences might provide resistance to specific dsRNA constructs. This might be avoided by mixing different constructs for the same mRNA. But, a more general mechanism of resistance, such as evolution of increased degradation of dsRNA constructs could cause this technology to fail for all RNAi targets in such a weed. Stewardship of this technology will be extremely important if it is to last.

There are other technologies that can give a crop an advantage over weeds. For example, engineering phosphite metabolism into crops so that they can transform it to phosphate can eliminate the need to use phosphate fertilizers (López-Arredondo and Herrera-Estrella 2012). Phosphite rates that provide good phosphate levels to plants transformed with the *ptxD* gene of *Pseudomonas stutzeri* WM88 are toxic to plants without the gene (Fig. 7). So, phosphite can be used to simultaneously provide phosphorus to the crop and act as a non-selective herbicide to weeds. Before this work, phosphite had been proposed as a broad spectrum herbicide (Ruthbaum and Ballie 1964; McDonald et al. 2001), but its toxicity to crops was a major disadvantage. Two other potential advantages to this technology is that it would reduce water contamination with phosphate, and it is both toxic to some plant pathogens and induces plant defense systems against pathogens (McDonald et al. 2001; Pinto et al. 2012; Saindrenan et al. 1988). This attractive approach to weed management is being field tested. To avoid resistance to phosphite from evolving, such a technology should be combined with other weed management methods.

This short discussion of where we might go from here indicates that there are many weed management technologies that might be utilized in the future. Our experience with evolution of resistance to synthetic herbicides

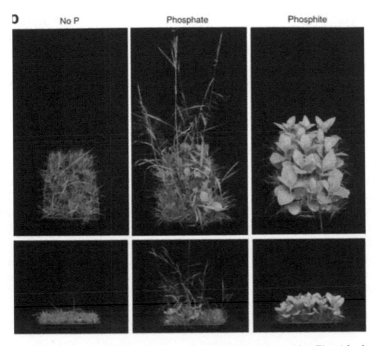

Figure 7. *Brachypodium distachyon* grown alone without phosphorus (no P), with phosphate and tobacco with a gene for converting phosphite to phosphate (Phosphate), and with the transgenic tobacco fertilized with phosphite (Phosphite) (from López-Arredondo and Herrera-Estrella 2012).

has made it clear that whatever technologies that are utilized, reliance on a single technology, no matter how attractive it may be, is unwise. So, future weed management should utilize multiple technologies in time and space to avoid loss of technologies by evolved resistance. This means that weed management will be more complicated and will require that farmers have more training or acquire expert advice to control weeds in a sustainable fashion. This scenario favors farmers with larger holdings, who can afford the cost of such advice. The problem of providing incentives to those who rent the land they farm to adopt weed management systems that prevent evolution of resistance is a difficult one that must be solved for resistance management to work on a broad scale.

Conclusion

Is the age of herbicides coming to an end in the near future? Herbicide use occupies a tiny time span in the millennia of cropping agriculture. So far, this small slice of time almost exactly encompasses the average human lifespan in developing countries. Just as antibiotic resistance has reached a crisis point in medicine, herbicide resistance has become a major issue in agronomic agriculture. In both cases, this predicament has not caused serious consideration of the abandonment of chemical solutions to deal with unwanted organisms. At this time and for the foreseeable future, weed management approaches that employ herbicides remain more cost effective and efficient than other weed control technologies. Renewed and more sophisticated discovery efforts may yield herbicides with new MOAs. If so, we hope that the lessons of the past three decades would cause the agrochemical and crop biotechnology industries and farmers to steward any such product much more carefully than they have ever cared or been able to do with previous products. Our experience with single solutions to weed management, as was briefly successful with glyphosate and GR crops, should preclude this from being tried again, no matter how attractive the technology. Part of the stewarding strategy for new and existing herbicides should be the incorporation of non-herbicide technologies for weed management, providing highly diverse weed management that should preclude the rapid, Darwinian response of weeds that we have seen to herbicides over the past few decades.

LITERATURE CITED

Aly R, Cholakh H, Joel DM, Leibman D, Steinitz B, Zelcer A, Naglis A, Yarden O and Gal-On A (2009) Gene silencing of mannose 6-phosphate reductase in the parasitic weed *Orobanche aegyptiaca* through the production of homologous dsRNA sequences in the host plant. Plant Biotechnol. J. 7: 487–498.

Ahmad-Hamdani MS, Yu Q, Han H, Cawthray GR, Wang SF and Powles SB (2012) Herbicide resistance endowed by enhanced rates of herbicide metabolism in wild oat (*Avena* spp.). Weed Sci. 61: 55–62.

Arias RS, Dayan FE, Michel A, Howell J and Scheffler BE (2006) Characterization of a higher plant herbicide-resistant phytoene desaturase and its use as a selectable marker. Plant Biotechnol. J. 4: 263–273.

Baerson SR, Sánchez-Moreiras A, Pedrol-Bonjoch N, Schulz M, Kagan IA, Agarwal AK, Reigosa MJ and Duke SO (2005) Detoxification and transcriptome response in *Arabidopsis* seedlings exposed to the allelochemical benzoxazolin-2-(3H)-one. J. Biol. Chem. 280: 21867–21881.

Bajsa J, Pan Z, Dayan FE, Owens DK and Duke SO (2012) Validation of serine/threonine protein phosphatase as the herbicide target site of endothall. Pestic. Biochem. Physiol. 102: 38–44.

Bass C and Field LM (2011) Gene amplification and insecticide resistance. Pest Manag. Sci. 67: 886–890.

Beckie HJ (2006) Herbicide-resistant weeds: Management tactics and practices. Weed Technol. 20: 793–814.

Bell MS, Hager AG and Tranel PJ (2013) Multiple resistance to herbicides from four site-of-action groups in waterhemp (*Amaranthus tuberculatus*). Weed Sci. 61: 460–468.

Belz RG (2007) Allelopathy in crop/weed interactions—an update. Pest Manag. Sci. 63: 308–326.

Bi Y, Liu W, Guo W, Li L, Yuan G, Du L and Wang J (2015) Molecular basis of multiple resistance to ACCase- and ALS-inhibiting herbicides in *Alopecurus japonicus* from China. Pestic. Biochem. Physiol. dx.doi.org/10.1016/j.pestbp.2015.07.002.

Bradshaw LD, Padgette SR, Kimball SL and Wells BH (1997) Perspectives on glyphosate resistance. Weed Technol. 11: 189–198.

Callaway RM and Ridenour WM (2014) Novel weapons: invasive success and the evolution of competitive ability. Frontiers Ecol. Environ. 2: 436–443.

Dayan FE and Duke SO (2014) Natural compounds as next generation herbicides. Plant Physiol. 166: 1090–1105.

Dayan FE, Ferreira D, Wang Y-H, Khan IA, McInroy JA and Pan Z (2008) A pathogenic fungi diphenyl either phytotoxin targets plant enoyl (acyl carrier protein) reductase. Plant Physiol. 147: 1062–1071.

Dayan FE, Howell JL and Weidenhamer JD (2009) Dynamic root exudation of sorgoleone and its *in planta* mechanisms of action. J. Exp. Bot. 60: 2107–2117.

Dayan FE, Owens DK, Tranel PJ, Preston C and Duke SO (2014) Evolution of resistance to phytoene desaturase and protoporphyrinogen oxidase inhibitors—state of knowledge. Pest Manag. Sci. 70: 1358–1366.

Dayan FE, Rimando AM, Pan Z, Baerson SR, Gimsing A-L and Duke SO (2010) Molecules of interest: Sorgoleone. Phytochemistry 71: 1032–1039.

Délye A, Menchari Y, Guillemin J-P, Matéjicek A, Michel S, Camilleri C and Chauvel B (2007) Status of black grass (*Alopercurus myosuroides*) resistance to acetyl-coenzyme A carboxylase inhibitors in France. Weed Res. 47: 95–105.

Délye A, Menchari Y, Michel S, Cadet E and Le Corre V (2013) A new insight into arable weed adaptive evolution: mutations endowing herbicide resistance also affect germination dynamics and seedling emergence. Ann. Bot. 45: 939–947.

Devine MD (2005). Why are there not more herbicide-tolerant crops. Pest Manag. Sci. 61: 312–317.

Devine MD, Duke SO and Fedtke C (1993) Physiology of Herbicide Action. Prentice Hall, Englewood Cliffs, NJ, USA, 441 pp.

Devonshire AL and Field LM (1991) Gene amplification and insecticide resistance. Annu. Rev. Entomol. 36: 1–21.

Duke SO (1988) Glyphosate. pp. 71–116. *In*: Kearney PC and Kaufman DD (eds.). Herbicides—Chemistry, Degradation and Mode of Action. Vol. III. Marcel Dekker, Inc., New York.

Duke SO (2010) Glyphosate degradation in glyphosate-resistant and –susceptible crops and weeds. J. Agric. Food Chem. 59: 5835–5841.

Duke SO (2012) Why have no new herbicide modes of action appeared in recent years? Pest Manag. Sci. 68: 505–512.

Duke SO (2014) Biotechnology: Herbicide-resistant crops. pp. 94–116. *In*: Van Alfen N (ed.). Encyclopedia of Agriculture and Food Systems. Vol. 2. Elsevier, San Diego.

Duke SO (2015) Perspectives on transgenic, herbicide crops in the USA almost 20 years after introduction. Pest Manag. Sci. 71: 652–657.

Duke SO and Powles SB (2008) Glyphosate: A once in a century herbicide. Pest Manag. Sci. 64: 319–325.

Duke SO and Dayan FE (2015) Discovery of new herbicide modes of action with natural phytotoxins. Amer. Chem. Soc. Symp. Ser. 1204: 79–92.

Duke SO, Baerson SR and Rimando AM (2003) Herbicides: Glyphosate. *In*: Plimmer JR, Gammon DW and Ragsdale NN (eds.). Encyclopedia of Agrochemicals. John Wiley & Sons, New York. http://www.mrw.interscience.wiley.com/eoa/articles/agr119/frame.html.

Duke SO, Baerson SR and Gressel J (2009) Genomics and weeds: A synthesis. pp. 221–247. *In*: Stewart CN (ed.). Weedy and Invasive Plant Genomics. Blackwell Publishing, Singapore.

Duke SO, Owens DK and Dayan FE (2014) The growing need for biochemical bioherbicides. Amer. Chem. Soc. Symp. Ser. 1172: 31–43.

Fennimore SA, Hanson BD, Sosnoskie LM, Samtani JB, Datta A, Knezevic SZ and Siemens MC (2014) Field applications of automated weed control: Western hemisphere. pp. 151–169. *In*: Young SL and Pierce FJ (eds.). Automation: The Future of Weed Control in Cropping Systems. Springer Science+Business Media, Dordrectht.

Frisvold GB, Hurley TM and Mitchell PD (2009) Adoption of best management practices to control weed resistance by corn, cotton, and soybean growers. AgBioForum. 12: 370–381.

Fuerst EP, Arntzen CJ, Pfister K and Penner D (1986) Herbicide cross-resistance in triazine-resistant biotypes of four species. Weed Sci. 34: 344–353.

Gaines TA, Zhang W, Wang D, Bukun B, Chisholm ST, Shaner DL, Nissen SJ, Patzoldt WL, Tranel PJ, Culpepper AS, Grey TL, Webster TM, Vencill WK, Sammons RD, Jiang J, Preston C, Leach JE and Westra P (2010) Gene amplification confers glyphosate resistance in *Amaranthus palmerii*. Proc. Natl. Acad. USA 107: 1029–1034.

Gaines TA, Lorentz L, Figge A, Herrmann J, Maiwald F, Ott M-C, Han H, Busi R, Yu Q, Powles SB and Beffa R (2014) RNA-seq transcriptome analysis to identify genes involved in metabolism-based diclofop resistance in *Lolium rigidum*. The Plant J. 78: 865–876.

Gaj T, Gersbach CA and Barbas CF (2013) ZFN, TALEN, and CRISPR/Cas-based methods for genome editing. Trends Biotechnol. 31: 397–405.

Gardner SN, Gressel J and Mangel M (1998) A revolving dose strategy to delay the evolution of both quantitative vs. major monogene resistances to pesticides and drugs. Internat. J. Pest Manag. 44: 161–180.

Ge X, d'Avignon DA, Ackerman JJH and Sammons RD (2010) Rapid vacuolar sequestration: the horseweed glyphosate resistance mechanism. Pest Manag. Sci. 66: 345–348.

Gerwick BC (2010) Thirty years of herbicide discovery: surveying the past and contemplating the future. Agrow (Silver Jubilee Edition): VII–IX.

Green JM (2014) Current state of herbicides in herbicide-resistant crops. Pest Manag. Sci. 70: 1351–1357.

Gressel J (2011) Low pesticide rates may hasten the evolution of resistance by increasing mutation frequencies. Pest Manag. Sci. 67: 253–257.

Harrington PM, Singh BK, Szamosi IT and Birk JH (1995) Synthesis and herbicidal activity of cyperin. J. Agric. Food Chem. 43: 804–808.

Heap I (2014) Global perspective of herbicide-resistant weeds. Pest Manag. Sci. 70: 1306–1315.

Heap I (2015) International Survey of Herbicide Resistant Weeds. http://weedscience.org/ - accessed Oct. 9, 2015.

Herbicide Resistance Action Committee: http://www.hracglobal.com/pages/classificationofherbicidesiteofaction.aspx accessed Oct. 19, 2015.

Hilton HW (1957) Herbicide tolerant strains of weeds. Hawaiian Sugar Plant Assoc. Annu. Rep. 69.

Huffman J, Hausman NE, Hager AG, Riechers DE and Tranel PJ (2015) Genetics and inheritance of nontarget-site resistance to atrazine and mesotrione in a waterhemp (*Amaranthus tuberculatus*) population from Illinois. Weed Sci. 799–809.

James C (2014) Global Status of Commercialized Biotech/GM Crops. International Service for the Acquisition of Agri-Biotech Applications. ISAAA Brief No. 49, Ithaca, NY.

Jander G, Baerson SR, Hudak JA, Gonzalez KA, Gruys KJ and Last RL (2003) Ethylmethanefulfonate saturation mutagenesis in *Arabidopsis* to determine the frequency of herbicide resistance. Plant Physiol. 131: 139–146.

Jeschke P (2016) Progress of modern agricultural chemistry and future prospects. Pest Manag. Sci. 72: DOI: 10.1002/ps.4190.

Jugulam M, Niehaus K, Godar AS, Koo D-H, Danilova T, Friebe B, Sehgal S, Varanasi VK, Wiersma A, Westra P, Stahlman PW and Gill BS (2014) Tandem amplification of chromosomal segment harboring 5-enolpyruvylshikimiate-3-phosphate synthase locus confers glyphosate resistance in Kochia scoparia. Plant Physiol. 166: 1200–1207.

Kaundun SS (2014) Resistance to acetyl-CoA carboxylase-inhibiting herbicides. Pest Manag. Sci. 70: 1405–1417.

Lee LJ and Ngim J (2000) A first report of glyphosate-resistant goosegrass (*Eleusine indica* (L.) Gaertn) in Malaysia. Pest Manag. Sci. 56: 336–339.

Lemmens M, Scholz U, Berthiller F, Dall'Asta C, Koutnik A, Schuhmacher R, Adam G, Buerstmayr H, Mesterházy Á, Krska R and Ruckenbauer P (2005) The ability to detoxify the mycotoxin deoxynivalenol colocalizes with a major quantitative trait locus for *Fusarium* head blight resistance in wheat. Molec. Plant-Microbe. Interact. 18: 1318–1324.

López-Arredondo DL and Herrera-Estrella L (2012) Engineering phosphorus metabolism in plants to produce a dual fertilization and weed control system. Nature Biotechnol. 30: 889–892.

Mallory-Smith CA, Thill DC and Dial MJ (1990) Identification of sulfonylurea herbicide-resistant prickly lettuce (*Lactuca serriola*). Weed Technol. 4: 163–168.

McDonald AE, Grant BR and Plaxton WC (2001) Phosphite (phophoric acid): its relevance in the environment and agriculture and influence on plant phosphate starvation response. J. Plant Nutr. 24: 1505–1519.

Nandula VK, Ray JD, Ribeiro DN, Pan Z and Reddy KN (2013) Glyphosate resistance in tall waterhemp (*Amaranthus tuberculatus*) from Mississippi is due to both altered target-site and nontarget-site mechanisms. Weed Sci. 61: 374–383.

Norsworthy JK, Ward SM, Shaw DR, Llewellyn RS, Nichols RL, Webster TM, Bradley KW, Frisvold G, Powles SB, Burgos NR, Witt WW and Barrett M (2012) Reducing risks of herbicide resistance: Best management practices and recommendations. Weed Sci. 60 (sp1): 31–62.

Olson S (2015) An analysis of the biopesticide market now and where it is going. Outlooks Pest Manag. 26: 203–206.

Patzoldt WL, Hager AG, McCormick JS and Tranel PJ (2006) A codon deletion confers resistance to herbicides inhibiting protoporphyrinogen oxidase. Proc. Natl. Acad. Sci. USA 103: 12329–12334.

Pinto KMS, do Nascimento LC, Gomes ECD, da Silva HF and Miranda JD (2012) Efficiency of resistance elicitors in management of grapevine downy mildew *Plasmopara vitcola*: epidemiological, biochemical and economic aspects. Eur. J. Plant Pathol. 134: 745–754.

Powles SB, Preston C, Bryan IB and Jutsum AR (1997) Herbicide resistance: Impact and management. Adv. Agron. 58: 57–93.

Powles SB, Lorraine-Colwill DF, Dellow JJ and Preston C (1998) Evolved resistance to glyphosate in rigid ryegrass (*Lolium rigidum*) in Australia. Weed Sci. 46: 604–607.

Pratley JE, Urwin NAR, Stanton RA, Baines PR, Broster JC, Cullis K, Schafer DE, Bohn JA and Kruger RW (1999) Resistance to glyphosate in *Lolium rigidum*. I. Bioevaluation. Weed Sci. 47: 405–411.

Reddy KN, Rimando AM and Duke SO (2004) Aminomethylphosphonic acid, a metabolite of glyphosate, causes injury in glyphosate-treated, glyphosate-resistant soybean. J. Agric. Food Chem. 52: 5139–5143.

Reddy KN, Rimando AM, Duke SO and Nandula VK (2008) Aminomethylphosphonic acid accumulation in plant species treated with glyphosate. J. Agric. Food Chem. 56: 2125–2130.

Ruthbaum HP and Ballie WJH (1964) The use of red phosphorous as a fertilizer. Part 4. Phosphite and phosphate retention in the soil. N Z J. Sci. 7: 446–451.

Ryan GF (1970) Resistance of common groundsel to simazine and atrazine. Weed Sci. 18: 614–616.

Saindrenan P, Barchietto T and Gilbert B (1988) Modification of the phosphite induced resistance response in leaves of cowpea infected with *Phytophthora cryptogea* by α-aminooxyacetate. Plant Sci. 58: 245–252.

Salas RA, Dayan FE, Pan Z, Watson SB, Dickson JW, Scott RC and Burgos NR (2012) EPSPS gene amplification in glyphosate-resistant Italian ryegrass (*Lolium perenne* ssp. *multiflorum*) from Arkansas. Pest Manag. Sci. 68: 1223–1230.

Sammons RD and Gaines TA (2014) Glyphosate resistance: State of knowledge. Pest Manag. Sci. 70: 1367–1377.

Sauer NJ, Mozoruk J, Miller RB, Warburg ZJ, Walker KA, Beetham PR, Schöpke CR and Gocal GFW (2015) Oligonucleotide-directed mutagenesis for precision gene editing. Plant Biotechnol. J. DOI: 10.1111/pbi.12496.

Schulz M, Marocco A, Tabaglio V, Macias FA and Molinillo JMG (2013) Benzoxazinoids in rye allelopathy—from discovery to application in sustainable weed control and organic farming. J. Chem. Ecol. 39: 154–174.

Shaner DL (2000) The impact of glyphosate-tolerant crops on the use of other herbicides and on resistance management. Pest Manag. Sci. 56: 320–326.

Shaner DL and Beckie HJ (2014) The future of weed control and technology. Pest Manag. Sci. 70: 1329–1339.

Svitashev S, Young JK, Schwartz C, Gao H, Falco SC and Cigan AM (2015) Targeted mutagenesis, precise gene editing, and site-specific gene insertion in maize using Cas9 and guide RNA. Plant Physiol. 169: 931–945.

Switzer CM (1957) The existence of 2,4-D resistant strains of wild carrot. Proc. Northeast Weed Control Conf. 11: 315–318.

Tan S, Evans RR, Dahmer ML, Singh BK and Shaner DL (2005) Imidazolinone-tolerant crops: History, current status and future. Pest Manag. Sci. 61: 246–257.

Timmons FL (2005) A history of weed control in the United States and Canada. Weed Sci. 53: 748–761.

Tranel PJ and Wright TR (2002) Resistance of weeds to ALS-inhibiting herbicides: What have we learned? Weed Sci. 50: 700–712.

US Department of Agriculture: https://www.aphis.usda.gov/biotechnology/petitions_table_pending.shtml). Accessed Nov. 8, 2015.

Voytas DF and Gao C (2014) Precision genome engineering and agriculture: Opportunities and regulatory challenges. PLoS Biol. 12(6): e1001877. Doi:10.1371/journal.pbio.1001877.

Yu Q and Powles SB (2014a) Resistance to AHAS inhibitor herbicides: current understanding. Pest Manag. Sci. 70: 1340–1350.

Yu Q and Powles SB (2014b) Metabolism-based herbicide resistance and cross-resistance in crop weeds: A threat to herbicide sustainability and global crop production. Plant Physiol. 166: 1106–1118.

Yu Q, Han H, Cawthray GR, Wang SF and Powles SB (2013) Enhanced rates of herbicide metabolism in low herbicide-dose selected resistant *Lolium rigidum*. Plant Cell Environ. 36: 818–827.

Yu Q, Jalaludin A, Han H, Chen M, Sammons RD and Powles SB (2015) Evolution of a double amino acid substitution in the EPSP synthase in *Eleusine indica* conferring high level glyphosate resistance. Plant Physiol. doi: http://dx.doi.org/10.1104/pp.15.00146.

Chapter 6

Inter-specific Gene Flow from Herbicide-Tolerant Crops to their Wild Relatives

Amit J. Jhala,[1],* Debalin Sarangi,[1] Parminder Chahal,[1]
Ashish Saxena,[2] Muthukumar Bagavathiannan,[3]
Bhagirath Singh Chauhan[4] and Prashant Jha[5]

Introduction

The global policy agenda of agriculture is based around producing sufficient food, feed, and fiber to meet the needs of the world's growing population (Anonymous 2009). Additionally, agricultural cropping systems have been challenged to create oil-based fuels and plant-based bio-products, reduce use of pesticides, and develop crops tolerant to biotic and abiotic stresses. Therefore, crop cultivars with higher yields and resistance to important pests are required to meet food requirements in a sustainable manner. Technological advances for developing bio-products, improved crop cultivars, and pest and/or herbicide-tolerant crops are predicated on the use of plant biotechnology.

[1] Department of Agronomy and Horticulture, University of Nebraska-Lincoln, Lincoln, NE 68583.
[2] Dow AgroSciences, York, NE.
[3] Department of Soil and Crop Sciences, Texas A & M University, College Station, TX 77843.
[4] Queenland Alliance for Agriculture and Food Innovation (QAAFI), The University of Queensland, Toowoomba, Queensland, 4350, Australia.
[5] Southern Agricultural Research Center, Montana State University, Huntley, MT 59037.
* Corresponding author: Amit.Jhala@unl.edu

Gene flow is a natural phenomenon and may occur via pollen, seed, or vegetative propagules (Knispel et al. 2008; Mallory-Smith and Zapiola 2008). Gene flow via pollen and seed from glyphosate-resistant crops, such as canola (*Brassica napus* L.) and creeping bentgrass (*Agrostis stolonifera* L.) is evident (Mallory-Smith and Zapiola 2008). The commercialization of herbicide-tolerant crops in the late 1990s was controversial; however, this new technology has provided many benefits to growers (Beckie et al. 2006; Duke 2005). Besides the benefits of herbicide-tolerant crops, their widespread adoption has created controversies about the consequences of herbicide-tolerant crops and their potential negative impacts. The relevance of assessing weedy characteristics when considering the invasiveness of herbicide-tolerant crops has been the subject of much debate (Ellstrand and Schierenbeck 2006; Jhala et al. 2011), since gene flow from transgenic crops to their wild relatives may inadvertently cause wild plants to acquire traits that improve their fitness (Ellstrand 2003; Warwick et al. 2009). The mere presence of wild relatives in a given area does not necessarily imply that interspecific gene flow will occur, though the long-term co-existence of herbicide-tolerant crops and their wild relatives in a given habitat may signal the need to assess the likelihood of spontaneous gene transfer from the former to the latter (Jhala et al. 2008). For example, jointed goatgrass (*Aegilops cylindrical* Host.), a wild relative of wheat (*Triticum aestivum* L.) is capable of acquiring the herbicide-tolerant trait from wheat (Hanson et al. 2005). Similarly, wild and weedy sorghum occurring within and adjacent to cultivated sorghum (*Sorghum bicolor* L.) may pose risks for inter-specific transgene movement. The concern that the introgression of the herbicide-tolerant transgene into weedy rice (*Oryza sativa* L.) could result in a more difficult to manage hybrid is a serious consideration for the commercial production of novel rice in regions where rice and weedy rice co-exist (Chen et al. 2004). Due to these concerns, there was an increased interest in inter-specific hybridization between crops and their wild relatives after the commercialization of herbicide-tolerant crops (Jhala and Hall 2013), and therefore, information is needed on the geography and reproductive biology of the most important cultivated crops and their wild relatives along with their potential for inter-specific gene flow. Introgression involves the permanent transfer of genetic material from one plant to another, and may occur through repeated back-crossing. This chapter explains the center of origin of many common crop species, the occurrence and distribution of their wild relatives, their herbicide tolerant traits, and their potential for inter-specific gene flow.

Alfalfa

Alfalfa (*Medicago sativa* L.) is popularly known as the "Queen of forages" and is an important forage crop worldwide. In the United States, alfalfa is the fourth most widely grown crop behind corn (*Zea mays* L.), wheat, and soybean [*Glycine max* (L.) Merrill]. Alfalfa has excellent nutritional value as animal feed, with about 15 to 22% crude protein and is a good source of vitamins and minerals (Hanson et al. 1988). Because alfalfa is a leguminous crop, it provides soil health benefits to pasture lands when used in mixtures. Alfalfa can be used for direct grazing, cut and sold as hay, or processed into meal and cubes for export. It is predominantly used as feed for dairy cows, but is also an important feed for beef cattle, horses, chickens, turkeys, and other farm animals (NAAIC 2000). Additionally, alfalfa seed sprouts are directly consumed by humans and alfalfa juice is a component of some health products. Alfalfa has been considered as a candidate species for genetic modification with some important agronomic and quality traits, and alfalfa with novel traits such as glyphosate resistance and low-lignin content have already been commercialized in the United States.

Alfalfa originated at the Near Eastern Center, which includes Asia Minor, Iran, Transcaucasia, and Turkmenistan (Bolton 1962), although the exact location of origin is still under debate (e.g., Klinkowski 1933). It was the first domesticated forage crop and there are references to its use in Babylonia from 2650 B.P. (Bolton et al. 1972). Maritime trade connected by the Mediterranean region is believed to have helped alfalfa spread to regions further from its center of origin. Alfalfa was introduced to South and North America in the 15th century and to Australia and New Zealand in the 18th century (Barnes et al. 1988). It is widely adapted to various environmental conditions and is currently grown throughout the world.

The cultivated *Medicago sativa* complex includes *M. sativa* ssp. *sativa* (purple-flowered alfalfa), *M. sativa* ssp. *falcata* (yellow-flowered alfalfa), and *M. sativa* ssp. *varia* (variegated forms). These members of the *M. sativa* complex are known to occur outside of cultivation as feral populations (e.g., Bagavathiannan et al. 2010). Other subspecies in the *M. sativa* complex include *caerulea*, *glutinosa*, *tunetana*, *polychroa*, and *hemicycla* (Quiros and Bauchan 1988). *M. sativa* is known to hybridize among different subspecies within the complex (intra-specific) and also to some other species within the genus (inter-specific). The genus *Medicago* is comprised of 83 species, of which one third is perennials, including the cultivated *M. sativa*. Natural hybridization between the annual and perennial *Medicago* species is rare. Successful hybridization between *M. sativa* and its wild relatives has been reported for *M. glomerata*, *M. prostrata*, *M. cancellata*, *M. rhodopea*, *M. rupestris*, *M. saxatilis*, *M. daghestanica*, *M. pironae*, *M. dzhawakhetica*, *M. papillosa*, *M. marina*, *M. hybrida*, *M. arborea*, and *M. scutellata* (reviewed in Bagavathiannan and Van Acker 2009). In North America, the most common wild relative of alfalfa is *M. lupulina* (black medick), though hybridization between *M. sativa* and *M. lupulina* is unlikely.

Alfalfa is predominantly an insect-pollinated crop because "tripping" is required to release pollen and the presence of a strong stigmatic cuticle prevents self-pollination (Viands et al. 1988). Pollinators such as honey bees (*Apis mellifera*), leaf cutter bees (*Megachile rotundata*), alkali bees (*Nomia melander*), and bumble bees (*Bombus* spp.) are known to be important pollinators for alfalfa, but the distance of gene flow will depend on the foraging behavior of the specific pollinator. Alkali bees and honey bees are known to fly for several miles, whereas leaf cutter bees are short-distance foragers. The probability of gene flow is also dependent on the production system in question, whether hay or seed. Under field conditions, gene flow could occur from hay to hay, hay to seed, seed to hay, seed to seed, feral to hay, hay to feral, feral to seed, and seed to feral systems (Van Deynze et al. 2008).

A summary of experiments conducted by Putnam (2006), and Teuber et al. (2004, 2007) showed that seed to seed gene flow levels under honey bee pollination were under 3% at an isolation distance of 152 m and declined to < 0.5% at 325 m. These levels were much lower when leaf cutter bees were used for pollination. St. Amand et al. (2000) detected gene flow between alfalfa seed production fields and feral alfalfa populations at up to 1000 m, while Bagavathiannan et al. (2011) have estimated an outcrossing rate ranging from 62 and 85% between cultivated alfalfa seed production fields stocked with leaf cutter bees and adjacent feral alfalfa populations located along roadsides within a maximum study distance of 15 m. Seed-mediated gene flow is also likely to occur in alfalfa because of its small seed size—which makes it susceptible to spillage during transport—and its high nutritional status, since it can be consumed and dispersed by seed predators.

Barley

Cultivated barley (*Hordeum vulgare* L. ssp. *vulgare*) is one of the founder crops of Neolithic food production in old world agriculture (Badr et al. 2000; Dai et al. 2012; Zohary et al. 2012). It is the fourth most important cereal grain in the world after corn, rice (*Oryza sativa* L.), and wheat, with a total production of more than 140 M metric tons worldwide (FAOSTAT 2013). Barley is a cool-season annual crop grown in the spring or winter. Historically it was a major source of human nutrition, but the increasing use of corn, rice, and wheat later restricted its use primarily to a feed, malting, and brewing grain (Baik and Ullrich 2008; Newton et al. 2011; Sullivan et al. 2013). There are no varieties of genetically modified (GM) barley available commercially, but GM traits for pest and disease resistance, biotic and abiotic stress tolerance, and improved malting and brewing qualities in barley are in development (GMO Compass 2010; Mrízová et al. 2014).

Though the northeast region of the Fertile Crescent (the Israel-Jordan area) is well known as the primary center of origin, diversity, and domestication for barley (Badr et al. 2000; Diamond 1999; Vavilov

1926; Zohary et al. 2012), there are many contradictory theories regarding barley's center of origin: various biologists consider the Himalayas (Tibet), Ethiopia, and Morocco as the centers of barley domestication (Dai et al. 2012; Molina-Cano et al. 1987; Ren et al. 2013). Harlan and Zohary (1966) noted that cereal domestication took place around 7000 B.C., giving major importance to barley and wheat in southwest Asia. However, a wild progenitor of cultivated barley (*H. vulgare* ssp. *vulgare*) is *H. vulgare* ssp. *spontaneum* (C. Koch) Thell., which can still be found in the Fertile Crescent from Israel and Jordan to southern Turkey, the Kurdistan portion of Iraq, and in southwestern Iran, provides evidence to support the Fertile Crescent region as the center of origin for barley diversification (Harlan and Zohary 1966; Jakob et al. 2014; Ma et al. 2012). Archaeological remains and molecular approaches have also confirmed the colonization of *H. spontaneum* in this area, as well as in the central Asian and Himalayan regions (Badr and El-Shazly 2012; Badr et al. 2000; Dai et al. 2012). The genus *Hordeum* can now be found in most temperate areas, including the subtropical regions of Central and South America and the arctic zones of central Asia and North America (Andersson and de Vincente 2010; von Bothmer and Komatsuda 2011).

Hordeum comprises about 30 annual and perennial species and is part of the tribe Triticeae (von Bothmer and Jacobsen 1985; Doebley et al. 1992). The genus *Hordeum* has a basic chromosome number $x = 7$, with both diploid ($2n = 2x = 14$) and polyploid ($2n = 4x = 28$; and $2n = 6x = 42$) species known to exist (Komatsuda et al. 1999). Cultivated two-rowed barley, *H. vulgare* ssp. *vulgare* and its wild progenitor *H. vulgare* ssp. *spontaneum* are diploid species. The morphological similarities, crosses, and back crosses between these two taxa led to their being treated as subspecies of *H. vulgare* rather than separate species (von Bothmer and Komatsuda 2011; Harlan and Zohary 1966). Von Bothmer and Jacobsen (1985) also considered *Hordeum bulbosum* and *Hordeum murinum* as closely related species to *H. vulgare* due to their morphological and ecological similarities, along with their high degree of chromosomal pairing at meiosis for the hybrid between *H. vulgare* and *H. bulbosum*.

Domesticated barley reproduces mainly by self-fertilization, causing lower natural outcrossing rates of $< 2\%$ at field level (Abdel-Ghani et al. 2004; Gatford et al. 2006; Parzies et al. 2000; Wagner and Allard 1991). Therefore, the European Environment Agency (EEA) has denoted barley as a "low-risk crop" for crop-to-crop and crop-to-wild gene flow (Eastham and Sweet 2002). Ritala et al. (2002) estimated 0 to 7% outcrossing frequency in barley varieties with an open-flowering habit at a distance of 1 m depending on the weather conditions. Brown et al. (1978) also estimated the chances of outcrossing in 26 wild barley populations in Israel and reported that the outcrossing varied from 0 to 9.6% with a mean of 1.6%. The isolation distance for barley seed production is smaller than that of many popular crops, varying between 0 to 3 m in the US and Canada, and 20 to 50 m in the OECD (Organization for Economic Cooperation and Development) countries and in Europe (Andersson and deVicente 2010).

Canola

Modern canola (*Brassica napus* L.) originated in the 1960s through the dedicated work of Canadian plant breeders who excluded the undesirable traits of rapeseed to develop a healthier oilseed crop. Canola is the second-largest oilseed crop worldwide after soybean, producing over 72 M metric tons (FAOSTAT 2013). China, India, Canada, France, Germany, and Australia are major canola-producing countries (Wittkop et al. 2009). Several recent canola cultivars have been released with a variety of health benefits, thus increasing market demand for canola oil.

The genus *Brassica* L. includes approximately 100 species, several of which are cultivated throughout the world. The majority of the Brassicas, including canola, originated in the Mediterranean region, though different species have specific centers of origin. Andersson and de Vicente (2010) suggested two prominent and independent centers of origin for canola in the north European Atlantic coast and the Afghanistan regions. Canola (previously known as rapeseed) was known to humans in ancient times and was consumed in Southeast Asia as early as 5,000 BC (Gupta and Pratap 2007), where it was used for medicinal purposes, edible oils, and oil-burning lamps. Rapeseed was also successfully domesticated and widely grown in many European countries around the period of 1600 AD. *Brassica napus* is a relatively recent Brassic aceae crop, the product of natural hybridization between *B. rapa* and *B. oleracea* (Nagaharu 1935; Chen et al. 2008).

Nagaharu (1935) presented the famous Brassica triangle that explains the complex relationship and origin of different species in the genus *Brassica* L. *Brassica napus* (AACC, n = 19) was created through a cross between *B. oleracea* (CC, n = 9) and *B. rapa* (AA, n = 10); *B. carnita* (BBCC, n = 17 was a product of *B. oleracea* (CC, n = 9) and *B. nigra* (BB, n = 8); while a cross between *B. nigra* (BB, n = 8) and *B. rapa* (AA, n = 10) resulted in *B. juncea* (AABB, n = 18). *B. napus* is an allotetraploid (2n = 38) as the result of a cross between two species, meaning that wild relatives of canola are consequently difficult to find. At the same time, many widely grown domesticated crops of *B. rapa*, *B. oleracea*, and *B. nigra* have an array of wild relatives mainly found in the Mediterranean region. Several Brassica species such as *B. deflexa*, *B. elongata*, *B. tournefortii*, and other related genera such as *Raphanus*, *Sinapis*, *Diplotaxis*, *Coincya*, *Moricandia*, *Rapistrum*, etc. are either cultivated or found in wild forms in many parts of the world and have the potential for outcrossing with *B. napus* (Raymer 2002; Andersson and Carmen de Vicente 2010).

Wind and insects are the primary sources of pollen dispersal in *B. napus*. As it is a partially allogamous crop with related wild and weedy species located throughout the world, gene flow is a serious concern (Warwick et al. 2003). Several studies conducted on gene flow reported outcrossing rates varying from 3 to 21% depending on wind conditions and the prevalence of insects (Anderson and de Vicente 2010). Honeybees, bumblebees, many native bees, and some beetles are known vectors of canola pollen, capable of transporting pollen grains anywhere from several meters to several kilometers (Mesquida et al. 1982; Cresswell 1999). The isolation distance for most canola foundation seed production sites is 200 or 400 m, depending on the required purity.

At present, many genetically-modified (GM) cultivars of canola are available for commercial production. In an experiment, Warwick et al. (2003) observed a high risk of pollen dispersal and hybridization between GM cultivars of canola and *B. rapa*; however, the chances of gene flow to other genera such as *Raphanus*, *Sinapis*, and *Erucastrum* are much lower. FitzJohn et al. (2007) provided a detailed account of the hybridization risk between the wild and domesticated relatives of Brassicas and their potential for transgenic escape. Meanwhile, Devos et al. (2009) conducted an extensive study to measure introgressive hybridization between canola and its wild relatives, observing that *B. napus* L. is highly susceptible to gene flow based on its introgressive hybridization propensity. Knispen et al. (2008) reported that pollen-mediated gene flow among glyphosate-, glufosinate-, and imidazolinone-resistant canola resulted in the evolution of multiple herbicide-resistant canola volunteers (Knispel et al. 2008). Despite the high propensity of *B. napus* to hybridization, most Brassicas are not able to successfully hybridize mainly due to ploidy barriers, and often require artificial means such as embryo rescue or ovary culture to create artificial hybridization.

Corn

Corn is cultivated in the tropical, subtropical, and temperate regions of the world. In terms of total area under cultivation and production, corn holds the third position worldwide following rice and wheat, and the United States is the largest corn-producing country in the world followed by China and Brazil (USDA 2015). The high-yielding hybrid corn planted today is the product of a single domestication event that occurred 6,000 to 10,000 years ago in Mexico through human selection of annual teosinte (*Zea mays* subsp. *parviglumis* H.H. Iltis and Doebley), with continuous selection and improvement thereafter (Matsuoka et al. 2002). The "Teosintes" are commonly referred to as a group of four annual and perennial species of the genus *Zea* native to a region ranging from northern Mexico to western parts of Nicaragua (Doebley 1990). Most of these species are genetically dissimilar to domesticated corn, except *Zea mays* ssp. *parviglumis,* which has a close genetic relationship with domesticated corn and is native to southwestern Mexico (Doebley 1990; Matsuoka et al. 2002).

The wild relatives of domesticated corn are divided into the genera *Zea* and *Tripsacum* (Doebley and Iltis 1980). The species of the genus *Zea* (teosintes) are further categorized into two sections: *Zea* and *Luxuriantes* (Doebley and Iltis 1980). The former section includes four species: *Zea mays* L. subsp. *mays* (domesticated corn), *Z. mays* subsp. *mexicana* H.H. Iltis (Mexican annual teosinte), *Z. mays* subsp. *parviglumis* H.H. Iltis and Doebley (annual teosinte), and *Z. mays* subsp. *huehuetenangensis* H.H. Iltis and Doebley (annual teosinte). There are four species in section *Tripsacum*: *Z. diploperennis* H.H. Iltis et al. (diploperennial teosinte), *Z. luxurians* (Durieu and Asch.) R.M. Bird (Guatemala or Florida teosinte), *Z. nicaraguensis* H.H. Iltis and B.F. Benz (Nicaraguan teosinte), and *Z. perennis* Reeves and Mangelsd (perennial teosinte). The second group of wild relatives of domesticated corn includes sixteen species in the genus *Tripsacum* L., and this group is further divided into two sections: *Fasciculata* Hitchc. and *Tripsacum* (de Wet et al. 1982, 1983).

The teosintes have the ability to shed seeds naturally, and these seeds have the ability to survive even inside the intestinal tracts of cattle (Wilkes 1977), enabling teosintes to disperse their seeds over longer distances and spring up as weeds in Mexican cornfields. For example, the dispersal of one teosinte species, *Z. mays* subsp. *mexicana,* has been reported in newly planted corn fields that were fertilized with cattle manure (Berthaud and Gepts 2004; Wilkes 1977).

Hybridization has been known to occur between corn and most of its wild relatives in the genus *Zea.* Naturally occurring hybrids between corn and the teosintes are often observed growing near or in Mexican cornfields (Wilkes 1977; Kermicle and Allen 1990), though hand-pollination of teosintes was less successful when corn was used as a pollen source, rather than vice versa (Kermicle 1997; Ting 1963; Kermicle and Allen 1990). Corn is not sexually compatible with most of the species of the genus *Tripsacum* and far fewer cases of hybridization are known to occur in the wild. However, artificial techniques have been used to successfully produce viable hybrids between corn and the *Tripsacum* species, though the resulting hybrids were mostly sterile (de Wet et al. 1973).

Corn pollen is dispersed predominantly by wind, and because of their large size, almost all of the pollen grains are dispersed within a 100 m radius of the pollen source (Raynor et al. 1972; Bannert et al. 2008). Outcrossing occurs at low rates between domesticated corn and its wild relatives (teosintes) in Mexico when teosintes act as pollen donors (Baltazar et al. 2005), which could be due to the isolation distance and non-synchronization of flowering between cultivated corn and its wild relatives in Mexico (Wilkes

1977). However, the pollen dispersal distance from the pollen source and the outcrossing frequency can vary depending on environmental factors, including wind direction, wind speed, temperature, and humidity (Ma et al. 2004). There is little scientific literature available regarding the outcrossing frequencies between corn and teosintes at different isolation distances when domesticated corn is used as a pollen donor. Even though the chances of outcrossing are very rare, there is the likelihood of gene flow via pollen flow from genetically modified (GM) corn cultivars to their wild relatives in Mexico and Central America.

Cotton

Cotton is an ancient fiber crop long known to humans. The genus *Gossypium* comprises about 50 species, of which four different species produce spinnable fibers: *Gossypium herbaceum* and *G. arboreum*, the diploid (2n = 26; AA genome) old world species; and *G. hirsutum* and *G. barbadense*, the autotetraploid (2n = 52; AADD genome) new world species (Smith and Cothren 1999). Cultivated *G. herbaceum* was domesticated from a naturally occurring progenitor *G. herbaceum* var *africanum* in southern Africa (Hutchinson 1959); however, *G. herbaceum* was also believed to have been domesticated in southern Arabia where its wild forms were introduced by traders (Smith and Cothren 1999). The site of domestication of *G. arboreum* is not clearly evident, but the Indus valley (present-day India and Pakistan) has been considered as one likely center of origin (Hutchinson 1959).

   G. hirsutum was domesticated in the Mesoamerican region (present-day Mexico). The native range of *G. hirsutum* includes the semiarid tropics and subtropics of the Caribbean, the northern regions of South America, and Central America, where the species occurs in a continuum of diverse wild and domesticated forms (Hutchinson 1951). Its wild forms are typically found in the undisturbed coastal regions and are morphologically distinct from its inland cultivated and feral forms. While there are debates regarding whether the coastal populations are truly wild or derivatives of the ferals (Hutchinson 1951; Fryxell 1979), the inland *G. hirsutum* populations are all associated with human settlements. Using nuclear restriction fragment-length polymorphism markers (RFLPs), Brubaker and Wendel (1994) traced the origin of *G. hirsutum* to southern Mexico and Guatemala. Currently, *G. hirsutum* (upland cotton) is the most widely grown cotton species in the world, with a production of 119 million bales (218 kg/bale) in 2014 (Cotton Incorporated 2015). Cotton is predominantly grown in the tropical and subtropical climatic regions, with India, China, and the United States as the top three producers of cotton in the world. *G. barbadense* (also known as pima cotton, Sea Island cotton, or Egyptian cotton) is thought to have originated in western South America and has been domesticated in the southern region of Ecuador (Westengen et al. 2005).

The remaining 46 species of cotton are distributed as wild forms throughout the tropics and subtropics. The majority of cultivated cotton's wild relatives are predominantly confined to their respective centers of origin. A detailed description and distribution of the various wild species of cotton is provided by Gotmare et al. (2015). In the United States, wild cotton relatives are found in some localized areas, including Hawaii, Florida, Puerto Rico, and the US Virgin Islands (Mendelsohn et al. 2003), though volunteer and feral forms of the cultivated species may be found in cotton-growing areas. In Texas, feral cotton is commonly observed along roadsides in the Rio Grande valley, where it could have been established through seed spillage during transportation (Bagavathiannan, personal observations).

Upland cotton produces heavy and sticky pollen that are not suitable for wind dispersal (McGregor 1976). Instead, cotton pollination is typically mediated by insects, especially different species of native bees (Moffett et al. 1975; Rao et al. 1996), since the plant produces large, attractive flowers that are adapted for insect pollination. The flowers of most wild cotton species are more striking than their domesticated relatives. Since stigma is receptive only for a single day from sunrise to midafternoon, pollination occurs within a short window (McGregor 1976), with pollen grain viability declining to zero within about 12 hrs after dehiscence (Richards et al. 2005).

Hybridization between compatible cotton relatives is likely and historical evidence also supports hybridization among various *Gossypium* lineages over time (Wegier et al. 2011). In upland cotton, outcrossing rates are usually below 10% and are typically affected by pollinator activity and environmental conditions, both of which vary by location and time (Richmond 1951; Kareiva et al. 1994; Van Deynze et al. 2005). In a study conducted by Yan et al. (2015), pollen-mediated gene flow was about 11% for *Bombus ignitus*, while it was only 3% with *Pieris rapae*. Van Deynze et al. (2005) documented an outcrossing rate of 7.65% at a 0.3 m distance under a high activity of *Apis mellifera* L., whereas outcrossing rates declined to < 1% beyond 9 m from the pollen source. In the absence of high pollinator activity, gene flow was < 1% beyond 1 m. A review of crop-wild hybridization in cotton was also provided by Andersson and de Vicente (2010). A meta-analysis conducted by Kareiva et al. (1994) investigating data pertaining to > 15,000 samples originating from Arizona, Arkansas, Mississippi, and North Carolina revealed that gene flow was < 1% beyond 10 m in all locations.

Cowpea

Cowpea [*Vigna unguiculata* (L.) Walp. subsp. *unguiculata* var. *unguiculata*] is an important grain legume and fodder crop in the tropical and subtropical regions of the world (Westphal 1974). Cowpea is grown primarily in Africa, Southeast Asia, Australia, Cuba, and India. Nigeria is the largest producer of cowpea in the world followed by Niger, Brazil, Haiti, and India (TJAI 2010). Domesticated cowpea originated in Africa, most likely from its wild relative *Vigna unguiculata* ssp. *unguiculata* var. *spontanea* (Schweinf.) Pasquet (Pasquet 1999); however, the actual location of cowpea domestication within Africa is unknown.

The species *Vigna unguiculata* includes domesticated cowpea, annual wild cowpea (*Vigna unguiculata* ssp. *unguiculata* var. *spontanea*), and ten perennial wild forms of cowpea (Pasquet 1993a, 1993b, 1997). The domesticated form of cowpea further includes five cultivar groups, cv. *unguiculata*, cv. *biflora*, cv. *sesquipedalis*, cv. *melanophthalmus*, cv. *textilis*. The wild forms of cowpea can be harvested as fodder crops, and as a result are not considered weeds in cultivated fields in Africa (Berville et al. 2005; Feleke et al. 2006).

Cowpea is a highly self-pollinated species with limited possibilities of outcrossing; however, outcrossing rates of up to 15% have been reported in domesticated cowpea (Duke 1981). Domesticated cowpea can hybridize with some of its wild relatives and can yield viable and fertile hybrids using hand pollination (Panella and Gepts 1992; Smartt 1990). However, the compatibility of a cross between the domesticated and wild forms decreases when domesticated cowpea serves as a pollen donor (Feleke et al. 2006). Domesticated and wild cowpeas (*V. unguiculata* ssp. *unguiculata* var. *spontanea*) have fodder value and sometimes grow together in the same field; as a result, hybrids between domesticated and wild cowpea are often observed in cultivated fields in West Africa (Feleke et al. 2006). Few genetic transformations have been attempted in cowpea, and no herbicide-tolerant cultivar of cowpea is available commercially. While little in the way of scientific literature is available on the topic, the possibility of outcrossing of cultivated cowpea with closely related wild species in Africa is limited under natural conditions.

Oat

Oat (*Avena sativa* L.) belongs to the genus *Avena*, which consists of around 7 to 34 diploid, tetraploid, and hexaploid species (Baum 1977; Ladizinsky and Zohary 1971). The most widely cultivated oat, also known as common oat (*Avena sativa* L.), is an allohexaploid (2n = 6x = 42) and is mainly self-pollinated with low rates of outcrossing. Cultivated hexaploid oats have their center of diversity in the Middle East and were domesticated within the last 4,000 years (Zohary and Hopaf 1988; Ladizinsky 1988). Prior to their domestication, oats existed as wild and weedy plants in wheat and barley fields in the Middle East. Nowadays, domesticated oats are important cereal crops in the world's temperate zones.

There is evidence of independent domestication at each ploidy level in the genus *Avena* (Harlan 1977); for example, *Avena strigosa*, a diploid oat species, was domesticated from the predecessor of *A. weistii* and *A. hirtula* in Europe. At the tertraploid level, a tetraploid species, *A. abyssinica*, was domesticated from the wild ancestor of *A. vaviloviana* in Ethiopia. Similarly, the hexaploid oats have also evolved from two independent domestications or hybridization events—with the first hybridization between two diploid species, and the second between the resulting tetraploid and another diploid species (Thomas 1995).

The wild oat species have the ability to shatter their seeds upon maturity, and these seeds remain dormant in the soil far longer than the domesticated forms, whose seeds are retained within the plants and lack dormancy (Thomas 1995). Wild oat species, with their seed shattering and dormancy characteristics, persist as weeds in cereal crops: for example, the wild relatives of domesticated oats, *A. fatua* L. and *A. sterilis* L., are noxious weed species in cereal crops throughout the world (Holm et al. 1977).

The genus *Avena* is divided into three groups (primary, secondary, and tertiary gene pools) with respect to *A. sativa* based on the ploidy levels (Legget and Thomas 1995). The primary gene pool consists of hexaploid species, including domesticated oats, and the hexaploid wild species in the primary pool are cross-compatible with domesticated oats. The secondary gene pool is comprised of tetraploid species, *A. maroccana* and *A. murphyi*, and these species have two genomes homologous with the cultivated species (Legget and Thomas 1995). A cross between the tetraploid and cultivated hexaploid species yields sterile pentaploid F1 hybrids; however, these F1 hybrids often produce seeds when crossed with either of the parents (Ladizinsky and Fainstein 1977). The tertiary gene pool includes the wild *Avena* species, which does not hybridize with common oat and requires artificial hybridization techniques such as embryo rescue to produce F1 hybrids. These wild species include tetraploid (*A. barbata* and *A. macrostachya*) and diploid species.

Previous studies have reported pollen-mediated gene flow frequencies between wild and domesticated oats where wild oats served as pollen donors. In one such study, Coffman and Weibe (1930) reported 0.96 to 6.6% outcrossing via pollen flow from wild to domesticated oats at a distance of 2.7 and 0.3 m from the pollen source, respectively. Similarly, Bickelmann and Leist (1988) observed a 0 to 6% outcrossing frequency through pollen transfer from wild to domesticated oats. Even though the outcrossing frequencies between wild and domesticated oats are very low, there is the likelihood of gene flow occurring in nature depending on the ploidy type and physical proximity of the plants. In fact, a hybrid between domesticated and hexaploid wild oat species was reported as a weed in some locations in Australia (Groves et al. 2000). No published literature is available, to our knowledge, regarding the frequency of outcrossing between domesticated and wild oat species when domesticated oat is used as the pollen donor;

however, the chances of gene introgression through pollen flow from domesticated oat to its wild relatives distributed throughout the Middle East cannot be neglected.

Pearl Millet

Pearl millet [*Pennisetum glaucum* (L.) R. Br.] is a resilient plant tolerant to adverse growing environments including drought, high temperatures, low fertility, high salinity, and low pH levels (Bidinger et al. 1987; Govindaraj et al. 2013; Kholová et al. 2010; Vadez et al. 2012). It is the major dietary constituent for people in arid and semi-arid tropical regions in Asia and Africa, and also has some fodder value (Clotault et al. 2012; Tako et al. 2015). India is the largest producer of pearl millet, where it is mostly grown in the hot and dry western parts of the country where the annual rainfall ranges between 300 to 800 mm (Vadez et al. 2012).

Pearl millet is the major cereal crop of the Sahel Zone of West Africa (Harlan et al. 1976; Brunken 1977). Though there are several centers of diversity reported for pearl millet, it is believed that pearl millet originated in the Sahel Zone (Oumar et al. 2008). Archaeobotanical evidence suggests that signs of pearl millet cultivation were found in the Mali region as early as 4,500 BP (Manning et al. 2011). India is considered the secondary center of diversity for pearl millet (Andersson and de Vicente 2010). Though the dry areas of Sahel are considered the primary center of origin, pearl millet is also cultivated in the more humid regions of southern Africa (Clotault et al. 2012).

Over 140 species of the genus *Pennisetum* are distributed throughout the world, and it is one of the largest genera in the tribe *Paniceae* (Brunken 1977). The genus is divided into five sections: *Brevivalvula* Döll, *Eupennisetum* Stapf, *Gymnothrix* (P. Beauv.) Benth. and Hook. f., *Heterostachya* Stapf and C.E. Hubb., and *Penicillaria* (Willd.) Benth. and Hook. f. (Stapf and Hubbard 1934). Pearl millet (*P. glaucum*) and its wild relatives are part of the sect. *Penicillaria*.

Cultivated pearl millet (*P. glaucum*), its wild relative (*P. violaceum*), and its weedy form (*P. sieberanum*) are all diploids ($2n = 2x = 14$). The wild species *P. violaceum* is considered the progenitor of pearl millet, while the weedy form (called shibras) is considered the intermediate form between cultivated pearl millet and its wild progenitor (Brunken et al. 1977). Shibras are closely associated with pearl millet and are common in western Africa and northern Namibia (Andersson and de Vicente 2010).

Pearl millet's primary gene pool (GP-1) includes the crop itself along with its wild and weedy relatives, and it is expected that all of these species can readily cross and produce at least a few fertile plants (Clayton and Renvoize 1982; Harlan and de Wet 1971; Stapf and Hubbard 1934). However, the secondary gene pool includes a tetraploid ($2n = 4x = 28$) perennial wild relative called elephant grass (*P. purpureum* Schumach.), which is considered one of the world's worst weeds (Brunken 1977).

Pearl millet is a hermaphroditic species with a possibility of cross-pollination up to 82% (Burton 1974; Nuijten and Richards 2011). Hybridization between domesticated pearl millet and its wild form, *P. violaceum*, is well-reported in several studies, and it is believed that their hybridization resulted in the evolution of *P. sieberanum* (Brunken et al. 1977; Mariac et al. 2006; Tostain 1992). Domesticated pearl millet and perennial elephant grass (*P. purpureum*) are sexually compatible, and hybridization has resulted in vigorous triploid ($2n = 3x = 21$) hybrids (Pantulu 1967; Techio et al. 2006). Gene flow between pearl millet and its wild relatives can be monitored by pre- and post-zygotic hybridization barriers such as pollen competition or reduced viability of hybrid grains (Robert et al. 1992; Sarr et al. 1988).

Leuck and Burton (1966) noted that wind is the most important factor for pollen dispersal. In a study in Rajasthan, India, vom Brocke et al. (2003) detected gene flow in pearl millet between two sites more than 100 km apart. In pearl millet, a 1 km isolation distance should be maintained for nucleus and breeder seed production, and a distance of at least 300 and 400 m should be maintained for certified and foundation seed production, respectively (Gupta 1999).

Rice

Rice is one of the world's most important crops, and is the primary food source for half of the world's population (Callaway 2014). Since the start of the "green revolution" in the 1960s–1970s, rice production has increased steadily due to the introduction of semi-dwarf and photoperiod insensitive rice varieties and the intense use of fertilizer, irrigation, and pesticides (Conway 1997; FAOSTAT 2013; Lu and Snow 2005). Global rice production totaled about 750 M metric tons in 2013, with > 90% of production share coming from Asia; China being the highest rice producer (≈ 200 M metric tons) followed by India (≈ 160 M metric tons) (FAOSTAT 2013). Cultivated rice is reported to be the first crop species in the world for which the complete genome has been sequenced, and it is considered to be a model plant for genetic research (IRGSP 2005; Lu and Snow 2005; Sasaki et al. 2002). Transgenic rice species resistant to herbicides and insects have been developed and are available on the market, while research is currently being conducted to develop transgenic rice tolerant to biotic and abiotic stress and to improve nutritional quality (ISAAA 2015; Lu and Snow 2005). "Golden rice," a well-known variety of transgenic rice, was developed by Ingo Potrykus and Peter Bayer to alleviate vitamin A deficiency in developing countries, including highly populated areas of Asia, Africa, and Latin America, and is under regulatory consideration to approve commercialization (Potrykus 2001; Ye et al. 2000).

Two cultivated rice species are grown throughout the world: Asian rice (*O. sativa*) and African rice (*O. glaberrima*). Asian cultivated rice, *O. sativa*, is grown throughout most of the rice producing regions, whereas African rice, *O. glaberrima*, is only grown on a limited scale in western and central Africa (Callaway 2014; Chang 1976; Khush 1997). The center of origin and domestication of cultivated rice (*O. sativa*) continues to be a lively debate that includes different locations such as the Ganges valley in northern India, the southern slope of the Himalayas, and different locations in China and southeast Asia (Andersson and de Vicente 2010; Callaway 2014). The genus *Oryza* is believed to have originated at least 130 million years ago on the supercontinent Gondwanaland, which later separated into the continents Antarctica, South America, Africa, and Australia, as well as parts of India and the Middle East (Chang 1976; Khush 1997). Initially, the Himalayan foothills or the Assam-Yunnan area (stretching from northeastern India to southwestern China) were thought to be the centers of domestication for Asian rice (Chang 1976; Watabe 1977); however, recent archaeological studies in China have revealed the center of origin to be the middle and lower basins of the Yangtze River (Callaway 2014; Fuller et al. 2007; Normile 2004; Sato 2000).

It is believed that rice was introduced to Greece and other Mediterranean countries by returning members of Alexander's expedition to India in 324 BC, and was later distributed through India to Madagascar, Africa, and Southeast Asia. Rice was most likely first introduced to the United States in 1685 from Madagascar (Khush 1997). The genus *Oryza* comprises two domesticated (*O. sativa*, Asian rice; *O. glaberrima*, African rice) and twenty-one wild species (Chang 1976; Khush 1997). Nine of the wild species are tetraploid ($2n = 4x = 48$) and the remaining wild and two cultivated species are diploid ($2n = 2x = 24$) (Ge et al. 1999; Khush 1997).

It is believed that the Asian wild grass *O. rufipogon* is the progenitor of cultivated rice, *O. sativa* (Callaway 2014; Oka 1974). The evolutions of Asian and African rice are considered to be the parallel independent evolutions of crop species (Chang 1976; Khush 1997; Second 1982). Weedy rice or red rice (*O. sativa* f. *spontanea*), a major problem weed in most rice-growing areas of the world, is believed to have evolved from the hybridization of *O. sativa* and *O. rufipogon* (Cao et al. 2006; Suh et al. 1997). Gene flow between cultivated rice and weedy rice is a great concern for the spread of transgenic traits (Busconi et al. 2012; Goulart et al. 2014).

The gene pool approach of Harlan and de Wet (1971) noted that the primary gene pool (GP-1) of cultivated Asian rice (*O. sativa*) encompasses both the ancestral wild species (*O. rufipogon* and *O. nivara*) and weedy rice

(*O. sativa* f. *spontanea*), so it is expected that these species can readily outcross and produce viable seeds. Cultivated rice is highly self-pollinated and an outcrossing percentage of less than 1% can be expected (Lu and Snow 2005; Rong et al. 2004). In an experiment conducted in Italy, Messeguer et al. (2001) revealed that the gene flow from transgenic rice to non-transgenic rice ranged from 0.01 to 0.53%. Though cultivated rice and its weedy relatives are primarily self-pollinated, low levels of crop-to-weed gene flow have been detected in several studies. Gene flow frequency from cultivated rice was estimated to be from 0.011 to 0.046% to weedy rice, and from 1.21 to 2.19% to the wild rice progenitor (Chen et al. 2004). Song et al. (2003) reported the highest crop-to-wild gene flow to be detected was < 3% in rice, with the maximum observed distance of gene flow being 43 m. Other studies reported that crop-weed hybrids have more vigor and similar or better fitness compared to their parent biotypes (Gealy et al. 2003; Langevein et al. 1990; Song et al. 2004). The isolation distance for commercial rice seed production is mostly less than 100 m, a distance that was determined by several regulatory authorities (Andersson and de Vicente 2010).

Sorghum

Sorghum [*Sorghum bicolor* (L.) Moench] is the fifth most important cereal crop in the world after corn, rice, wheat, and barley (Billot et al. 2013; FAOSTAT 2013). It is grown mostly in arid and semi-arid regions and has widely adapted to adverse environmental conditions (Hammer and Muchow 1994; Msongaleli et al. 2014). Sorghum is mostly grown for food, feed, fiber, and fuel (Amaducci et al. 2000; Dowling et al. 2002; Kim and Dale 2004; Schober et al. 2005; Zhao et al. 2008), and is the major staple food for a large number of people in the semi-arid and arid regions of Africa and Asia (Vara Prasad and Staggenborg 2010).

It is thought that sorghum originated in the northeast part of Africa around 5000–7000 years ago, with Ethiopia believed to be one of the centers of origin (de Wet and Harlan 1971; Kimber 2000; Vavilov 1951; Wendorf et al. 1992). Harlan (1971) also noted that savanna regions of the Sudan and Chad were probable areas of sorghum domestication. Sorghum was brought from Africa to the Middle East, China, and India at least 3,000 years ago through trading and shipping, and the Indian subcontinent is considered to be the secondary center of origin for sorghum (Dicko et al. 2006; Vara Prasad and Staggenborg 2010; Vavilov 1992). Later, in the 1700–1800s, sorghum was introduced to North America, and in the late nineteenth century it was introduced to Australia and South America (Dicko et al. 2006).

Sorghum belongs to the Poaceae (Gramineae) family and the tribe Andropogoneae. The genus *Sorghum* Moench is subdivided into five subgenera: *Chaetosorghum, Eu-Sorghum (Sorghum), Heterosorghum,*

Parasorghum, and *Stiposorghum*. The subgenera *Eu-Sorghum* or *Sorghum* includes cultivated grain sorghum, *S. bicolor* (L.) Moench ($2n = 2x = 20$), a complex of closely related annual weedy taxa from Africa; and a complex of perennial rhizomatous weedy taxa from Southern Europe and Asia, which includes *S. halepense* (L.) Pers. ($2n = 4x = 40$) and *S. propinquum* (Kunth.) Hitchc ($2n = 2x = 20$) (de Wet 1978; Ejeta and Grenier 2005). In the gene pool approach, Harlan and de Wet (1971) classified diploid *S. bicolor* and *S. propinquum* in the primary gene pool (GP-1), where the species can cross readily and produce fertile hybrids. However, the tetraploid *S. halepense* was classified under the secondary gene pool (GP-2), which includes species that can be crossed with GP-1 with reduced fertility in F_1; the gene transfer is possible but difficult. Other subgenera of the genus *Sorghum* were classified under the tertiary gene pool (GP-3), where hybrids with GP-1 will be anomalous, lethal, or completely sterile and gene transfer is not possible (Harlan and de Wet 1971; Ejeta and Grenier 2005). Additionally, Dweikat (2005) reported that the hybridization of diploid cultivated sorghum and tetraploid johnsongrass resulted in only one well-developed diploid seed out of 18,000 crosses, though most of the interspecific crosses were expected to produce sterile triploid F1 plants.

S. *bicolor* is divided into three subspecies, including *S. bicolor* ssp. *bicolor* that includes all domesticated grain sorghum, *S. bicolor* ssp. *arundinaceum* that includes the wild progenitor of cultivated sorghum, and *S. bicolor* ssp. *drummondii* that includes the hybrids among grain sorghum and its closest wild relatives (de Wet 1978). In a molecular analysis, Paterson et al. (1995) revealed that johnsongrass (*S. halepense*), the most widely recognized noxious weed in sorghum taxa, is a polyploid interspecific hybrid of *S. bicolor* × *S. propinquum*. Transgenic grain sorghum plants resistant to biotic-abiotic stress (thus improving grain quality) have been produced via genetic engineering (Visarada et al. 2014). Grain sorghum tolerant to nicosulfuron have been developed by DuPont Crop Protection in association with the Kansas State University Research Foundation, and are expected to be commercialized in the near future (Tuinstra and Al-Khatib 2008; Tuinstra and Al-Khatib 2010).

The crop-to-weed gene flow in sorghum is mostly dependent on several factors, including crossability, fertility, and fitness of the hybrids (Ejeta and Grenier 2005). Although sorghum is a self-pollinated species, the outcrossing rate can be up to 26% for grain sorghum with a compact panicle and > 60% for a species with an open panicle, such as sudangrass or *S. sudanense* (Piper) Stapf (Pedersen et al. 1998). The most widely recognized interspecies sorghum weed is shattercane (*S. bicolor* (L.) Moench ssp. *drummondii* (Nees ex Steud.) de Wet ex Davidse) (Defelice 2006). Schmidt et al. (2013) reported hybridization between cultivated sorghum and shattercane up to 16% at field level, and 2.6% of outcrossing at a distance of 200 m from the source or from the cultivated sorghum (Schmidt et al. 2013). Arriola and Ellstrand (1996) conducted an experiment to evaluate the outcrossing between crop sorghum and johnsongrass and measured 2% outcrossing at 100 m distance between grain sorghum and johnsongrass. In another study, Arriola and Ellstrand (1997) also evaluated the fitness of the crop-weed hybrids of grain sorghum and johnsongrass and concluded that the hybrid's vegetative and sexual characteristics were as fit as either parent. The isolation distances for sorghum seed production in the United States and OECD countries varies from 200 to 400 m depending on the seed categories (Andersson and de Vicente 2010).

Soybean

Soybean, a species native to eastern Asia, emerged as a domesticated crop around the eleventh century BC in northern and central China (Hymowitz 1990). Soybean was introduced to North America as a forage crop by Samuel Bowenin in 1765, and its prominence as a grain crop began in the 1920s (Hymowitz and Harlan 1983). Soybeans are grown in more than 90 countries worldwide, and the United States is the largest soybean producer in the world followed by Brazil and Argentina. These three countries together account for 82% of the world's total soybean production (SoyStats 2015).

Domesticated soybean is an annual tetraploid (2n = 40) of the family *Leguminosae*, subfamily *Papilionoideae*, genus *Glycine* Willd., and subgenus *Soja* (Moench). Domesticated soybean flowers are self-pollinated and highly autogamous, and plants exhibit limited natural outcrossing (< 1%) compared to their wild relatives (5 to 20%) (Fujita et al. 1997; Kiang et al. 1992; Poehlmann 1959; Woodworth 1922). The genus *Glycine* Willd. is divided into two subgenera: subg. *Soja* (Moench) F.J. Herm. and subg. *Glycine*. The subg. *Soja* includes domesticated soybean and its wild relatives, *G. soja* Sieb. and Zucc, and *G. gracilis* Skvortz (Skvortzow 1927). *G. soja* grows in fields, hedgerows, roadsides, and riverbanks in many Asian countries and is considered to be the wild ancestor of domesticated soybean. The subg. *Glycine* is comprised of around 15 of domesticated soybean's perennial wild relatives, most of which are native to Australia (Hymowitz et al. 1998).

Being a self-pollinated and highly autogamous plant species, the outcrossing rates within domesticated soybeans through pollen flow are very low (< 1%); however, the outcrossing rates may vary depending on the genotype, climatic conditions, and abundance of pollinators (Chiang and Kiang 1987; Palmer et al. 2001). Less than 1.5 and 0.1% outcrossing in domesticated soybeans was reported beyond distances of 1 and 2 m from the pollen source, respectively (Boerma and Moradshahi 1975); however, outcrossing frequencies of up to 20% have been reported in wild soybeans, much higher than the outcrossing rates for domesticated soybeans (Fujita et al. 1997).

Hybridization can occur naturally between domesticated soybean and its wild relatives in the subg. *Soja* (Oka 1983; Singh and Hymowitz 1989). A reduction in hybridization frequency was reported when domesticated soybean was used as a pollen donor compared to when its wild relatives were used as donors (Dorokhov et al. 2004; Singh and Hymowitz 1989). There is no scientific literature available, to our knowledge, regarding the outcrossing rates between domesticated and wild soybeans at different isolation distances with domesticated soybeans serving as the pollen donor. However, one study in Japan reported pollen-mediated gene flow < 1% from glyphosate-tolerant soybean to conventional soybean up to a distance of 7 m from the pollen source (Yoshimura et al. 2006). This suggests the distinct possibility of gene flow occurring via pollen from domesticated or GM soybean to its wild relatives (subg. *Soja*) in Asia, which is the center of wild soybean diversity.

Sunflower

The common sunflower (*Helianthus annuus* var. *annuus*) is a member of the Asteraceae or Compositae family of flowering plants. The center of origin of sunflower has been widely debated, though based on archaeological (Lentz et al. 2001) and linguistic (e.g., Lentz et al. 2008) evidence, Mexico has been suggested as the center of origin, despite a popular belief in North American origin. Recently, however, Blackman et al. (2011) reported evidence based on multiple evolutionary loci and neutral DNA markers that support a single domestication event in eastern North America.

The domesticated sunflower (*H. annuus* var. *macrocarpus*) has many wild relatives. The genus *Helianthus* is comprised of 52 species and 67 taxa (Schilling 2006). It has been suggested that wild *Helianthus* species grow in sympatry with domesticated sunflower in several locations (Schilling and Heiser 1981). An investigation by Kantar et al. (2015) revealed the range overlap of several wild *Helianthus* species with *H. annuus* in the US, which include *H. anomalus* (Utah, New Mexico), *H. argophyllus* (Texas), *H. arizonensis* (Arizona, New Mexico), *H. bolanderi* (California), *H. debilis* subsp. *cucmerifolius* (E. Texas), *H. deserticola* (Nevada, Utah, New Mexico), *H. divaricatus* (Central US), *H. exilis* (California), *H. grosseserratus* (Central US), *H. hirsutus* (Central US), *H. maximilliani* (Central US), *H. neglectus* (New Mexico), *H. niveus* subsp. *canescens* (California, Arizona, New Mexico), *H. niveus* subsp. *niveus* (Baja California), *H. niveus* subsp. *tephrodes* [California, Mexico (Sonora)], *H. paradoxus* (Texas, New Mexico), *H. pauciflorus* subsp. *pauciflorus* (Central US), *H. pauciflorus* subsp. *subrhomboideus* (Central US), *H. petiolaris* subsp. *fallax* (Western US), *H. petiolaris* subsp. *petiolaris* (Central US), *H. praecox* subsp. *hirtus* (West Texas), *H. praecox* subsp. *praecox* (East Texas), *H. praecox* subsp. *runyonii* (Texas), *H. salicifolius* (Oklahoma, Kansas, Arkansas,

Missouri), *H. silphioides* (Oklahoma, Arkansas, Missouri), *H. strumosus* (Central US), *H. tuberosus* (Central US), and *H. winteri* (California). Of these, *H. arizonensis*, *H. divaricatus*, *H. grosseserratus*, *H. hirsutus*, *H. maximilliani*, *H. niveus*, *H. pauciflorus*, *H. salicifolius*, *H. silphioides*, *H. strumosus*, *H. tuberosus*, and *H. winteri* show perennial growth characteristics. In agricultural areas, wild sunflowers are commonly found in cultivated fields and along roadside right-of-ways. In Texas, for instance, the wild forms have evolved into a major weed issue in row crops such as corn, cotton, and sorghum.

Although native to North America, *Helianthus* is widely adapted to areas outside of its native range. It was introduced to Europe, South America, Africa, Asia, and Australia. Currently, domesticated sunflower is an important oilseed crop worldwide, cultivated on about 43 million ha of land, with the majority of its production in Ukraine, Russia, the European Union, and Argentina (NSA 2015). Certain species of wild sunflower such as *H. tuberosus* (Jerusalem artichoke), *H. bolanderi* (serpentine sunflower), and *H. debilis* (cucumber leaf sunflower) were also introduced as ornamental species. In the introduced range, sunflower has often escaped cultivation and persisted as volunteer, feral, and weedy populations; for example, *H. annuus* ssp. *annuus* and *H. petiolaris* in Argentina (Poverene et al. 2004) and *H. argophyllus*, *H. bolanderi*, *H. petiolaris*, and *H. tuberosus* in Europe (Faure et al. 2002) have been found outside of cultivation as feral persistent populations. A detailed review of volunteer, feral, and weedy forms of sunflower was provided by Berville et al. (2005).

Sunflower is an erect growing plant with flower heads consisting of two types of flowers: outer ray flowers and inner disc flowers. The disc flowers are arranged in spiral whorls and mature progressively from the outer to the center (Weiss 1983). The anthers mature prior to stigma (protandrous), and flowers typically exhibit high levels of self-incompatibility, favoring cross-pollination. It is predominantly pollinated by insects, with honeybees, bumble bees, and solitary bees serving as important pollinators in the wild. Both domesticated and wild sunflower species show some degree of flowering synchrony, with an overlap in flowering between late May and early October in the US (Arias and Rieseberg 1994; Burke et al. 2002). A three-year survey conducted across the US by Burke et al. (2002) revealed that about two-thirds of cultivated sunflower fields occurred in close proximity to wild sunflower with extensive flowering synchrony.

Inter-specific hybridization has been widely reported in sunflower (Dedio and Putt 1980), and many of the wild taxa can hybridize well with crop sunflower (Chandler et al. 1986). Controlled crosses between cultivated and wild *H. annuus* produced fertile hybrids, whereas certain wider crosses with other *Helianthus* species have produced partially fertile hybrids (Heiser et al. 1969). Dorado et al. (1992) showed evidence that the majority of *H. petiolaris* from southern California have the cytoplasm of

H. annuus, suggesting the possibility of hybridization between the two species. Reports have documented a high degree of hybridization between cultivated and wild sunflower by as much as 42% near production fields (Linder et al. 1998; Whitton et al. 1997), and a large-scale survey by Burke et al. (2002) suggested hybridization in up to 33% of wild sunflower populations based on morphological evidence. Research by Snow et al. (1998) also documented crop-wild gene flow at a range of 5 to 40% within a distance of 1000 m from a cultivated sunflower field. In field experiments conducted in Kansas, gene flow from imidazolinone-resistant domesticated sunflower to wild relatives was detected up to 30 m from the pollen source, with a maximum gene flow of up to 22% at the 2.5 m distance (Massinga et al. 2003). In Argentina, Poverene et al. (2004) reported hybridization between *H. annuus* ssp. *annuus* and *H. petiolaris* at a mean frequency of 7%. Other studies not noted here have also reported the occurrence of hybridization between cultivated and wild sunflowers. Evidence further suggests that introgressed crop alleles may persist in wild sunflower populations for the long term, especially if the alleles provide neutral or favorable traits (Snow et al. 1998, 1999; Whitton et al. 1997).

Natural or induced mutations that confer resistance to ALS-inhibitor herbicides have been utilized in developing non-transgenic herbicide-tolerant sunflower lines (reviewed in Sala et al. 2012). In a Kansas soybean field, Al-Khatib et al. (1998) selected a weedy sunflower biotype with resistance to imazethapyr, an imidazolinone group herbicide. BASF commercialized this trait (known as ImiSun®) under its Clearfield™ production system. Further mutagenesis and selection of sunflower with imazapyr yielded the Clearfield Plus® (CLPlus) sunflower trait that surpassed ImiSun® in its level of herbicide tolerance. Al-Khatib et al. (1999) discovered a sulfonylurea-tolerant weedy sunflower population in eastern Kansas, which was introgressed into cultivated sunflower through forward crossing and selection with the herbicide tribenuron-methyl and is referred to as the "Sures" sunflower (Miller and Al-Khatib 2004). Gabard and Huby (2001) also selected the tribenuron-methyl tolerance trait through EMS mutagenesis of the sunflower line HA89. This trait was commercialized by DuPont™ Pioneer® under the name ExpressSun® sunflower, which is resistant to the herbicide Express® SG (Streit 2012).

Given the cross-compatibility between weedy and cultivated sunflower, pollen-mediated gene flow between the two is very likely to occur in several parts of the world. Thus, any novel trait placed in cultivated sunflower is capable of escaping into compatible wild and weedy sunflower populations present in agricultural landscapes. Crop-wild hybridization constitutes a primary risk, especially in the center of origin where the preservation of wild sunflower diversity would be critical for germplasm conservation and plant breeding (Canamutto and Poverene 2007). In southern Spain, feral populations of *H. annuus* ssp. *annuus* were found established in sunflower crop regions and can lead to intraspecific hybridization and the transfer of novel traits into the unmanaged environment (Berville et al. 2004). In other parts of Europe, feral populations of *H. tuberosus* and *H. annuus* could spread throughout the natural environment, although hybridization between the two typically results in sterile F1 derivatives (Faure et al. 2002). While it is unlikely that herbicide-tolerant traits will increase the weediness and invasiveness of sunflower (Sala et al. 2012), traits that provide adaptive advantage in a competitive environment such as abiotic stress tolerance may increase the persistence of the novel trait in the broader environment.

Wheat

Wheat is one of the oldest domesticated crops in human history, and many believe its domestication occurred at least 10,000 years ago. It remains one of the most important staple food crops worldwide, with an annual production of over 713 M metric tons (FAOSTAT 2013), trailing behind only corn and rice. China, India, the United States, Russia, France, Canada, and Germany rank as the top wheat-producing countries in the world. Wheat was one of the most important crops during the "Green Revolution" in the 1960s that almost tripled wheat productivity in a short period to save millions of lives throughout the developing world (Evenson and Gollin 2003; Waines and Ehdaie 2007). Until the last two decades, wheat was the largest grain crop in the world, though the success of hybrid corn and hybrid rice have since superseded global wheat production. There have been numerous attempts to utilize heterosis by producing hybrid wheat, but its adoption remains meager due to the higher cost of seed production. Because of wheat's complex ploidy levels and the public's opinion about genetically modified (GM) food (mainly bread), the adoption of GM technology in wheat is lagging behind that of other crops. "Clearfield" wheat is an herbicide-tolerant crop that combats the weed issues persisting in wheat, although Clearfield wheat was developed through a combination of mutant breeding and herbicide use rather than genetic modification (Johnson et al. 2002).

The Fertile Crescent in the Middle East has been widely accepted as the center of origin for modern bread wheat (Charmet 2011; McFadden and Sears 1946). Wheat belongs to the Poaceae or grass family, and has a wide variety of relatives with a multitude of ploidy levels. Bread wheat *T. aestivum* has six ploidy levels, a process that took thousands of years as diploid species crossed and developed first a tetraploid and finally a hexaploid species. Charmet (2011) observed based on historical data, archaeological surveys, and molecular marker studies conducted throughout the world that *T. aestivum* is the only hexaploid wheat-related domesticated species; nevertheless, its diploid and tetraploid progenitors have an abundance of wild relatives that also originated in the Fertile Crescent and were later distributed and developed in the Caucasian and Himalayan regions. The center of diversity for bread wheat is considered to be a narrow region along the eastern Mediterranean coast (Mac Key 2005), with a secondary center of diversity observed by Yamashita (1980) in the Hindu Kush Himalayan and Chinese regions.

A significantly large pool of Poaceae (grass family) members exist in domesticated forms, though the pool of wild relatives is much higher. Most wild relatives of wheat belong to the genus *Triticum*, *Aegilops*, *Agropyron*, *Dasypyrum*, *Elymus*, *Hordeum*, and *Secale*. The genus *Triticum* and *Aegilops* are primarily considered to be wheat species, and are grouped in the A, D, and U genome clusters (Kimber and Feldman 1987). Bread wheat (*Triticum aestivum* L.) was developed as the product of two unique and separate spontaneous hybridization events to produce a hexaploid (A genome cluster 2n = 6x = 42, AABBDD) crop (Allan 1980). The first primary spontaneous hybridization occurred between *T. urartu* (genome A) and *Aegilops speltoides* (genome B) to develop *T. turgidum* (tetraploid, genome AB) (Dvorak et al. 1993; Riley et al. 1958); later, a second spontaneous hybridization between a tetraploid *T. turgidum* (genome AB) and a diploid *Ae. tauschii* (genome D) was finally able to produce bread wheat *T. aestivum* (Warburton et al. 2006; Andersson and de Vicente 2010; Charmet 2011). Any wild relative comprised of either A, B or D genomes could potentially hybridize with bread wheat with a higher rate of success compared to wild relatives that lack homologous chromosomes (Andersson and de Vicente 2010). Based on sexual compatibility, three gene pools of bread wheat have been identified: primary (mainly diploid, tetraploid, and hexaploid *Triticum*; along with relatives comprised of the D genome); secondary (mostly of the genera *Aegilops*, *Secale*, and *Dasypyrum*); and tertiary (including species from the genera *Agropyron*, *Hordeum*, etc.).

The flowers of hexaploid *T. aestivum* produce relatively heavy pollen grains that have longevity of about half an hour, and because they lack nectar, insects are not a contributing factor to pollen dispersal. Wheat is predominantly an autogamous crop, and though wind could disperse pollen grains up to 30 m, this dispersal leads to no greater than 2% outcrossing (Redden 1977; Gatford et al. 2006). On occasion, however, some wheat cultivars have exhibited up to 10% outcrossing rates based on weather conditions and crop density (Waines and Hegde 2003; Andersson and de Vicente 2010). For instance, Hanson et al. (2005) found up to 8% hybridization between Clearfield imazamox-tolerant winter wheat (known as Clearfield wheat) and jointed goatgrass at distances up to 40 m. The jointed goatgrass-wheat hybrids were more tolerant to imazamox herbicide because they had a larger percentage of the *Als* 1 allele (confers tolerance to the herbicide) in their genome compared with the wheat-wheat hybrids.

In general, a 30 m isolation distance is considered effective for producing pure seeds; however, several wheat-producing countries have developed distancing norms between 3 and 50 m (Andersson and Carmen de Vicente 2010). In a commercial production field in Canada, Matus-Cadiz et al. (2007) observed trace levels of pollen gene flow at distances up to 2.75 km. There are currently multiple attempts to produce a GM wheat cultivar with resistance against diseases, pests, and herbicides, along with male sterility and improved quality traits, but as of now no true GM wheat cultivar is commercially available. Wheat's primary gene pool is most susceptible to gene flow based on its ease of hybridization with its wild relatives, and appropriate prevention measures should be taken to avoid any transgenic contamination in research studies.

LITERATURE CITED

Abdel-Ghani AH, Parzies HK, Omary A and Geiger HH (2004) Estimating the outcrossing rate of barley landraces and wild barley populations collected from ecologically different regions of Jordan. Theor. Appl. Genet. 109: 588–595.

Al-Khatib K, Baumgartner JR and Currie RS (1999) Survey of common sunflower (Helianthus annuus) resistance to ALS-inhibiting herbicides in northeast Kansas. pp. 210–215. *In*: Proceedings of the 21st Sunflower Research Workshop.

Al-Khatib K, Baumgartner JR, Peterson DE and Currie RS (1998) Imazethapyr resistance in common sunflower (*Helianthus annuus* L.). Weed Sci. 46: 403–407.

Allan RE (1980). Wheat. pp. 709–720. *In*: Fehr W and Hadley HH (eds.). Hybridization of Crop Plants. Madison, WI: American Society of Agronomy.

Amaducci S, Amaducci MT, Benati R and Venturi G (2000) Crop yield and quality parameters of four annual fibre crops (hemp, kenaf, maize and sorghum) in the North of Italy. Ind. Crops Prod. 11: 179–186.

Andersson MS and de Vicente MC (2010) Gene flow between crops and their wild relatives. Baltimore, MD: The Johns Hopkins University Press. pp. 49–69.

Anonymous (2009) United Nation's Millennium development Goals. United Nations, Washington DC.

Arias DM and Rieseberg LH (1994) Gene flow between cultivated and wild sunflowers. Theor. Appl. Genet. 89: 655–660.

Arriola PE and Ellstrand NC (1996) Crop-to-weed gene flow in the genus *Sorghum* (Poaceae): Spontaneous interspecific hybridization between johnsongrass, *Sorghum halepense*, and crop sorghum, *S. bicolor*. Am. J. Bot. 83: 1153–1160.

Arriola PE and Ellstrand NC (1997) Fitness of interspecific hybrids in the genus *Sorghum*: Persistence of crop genes in wild populations. Ecol. Appl. 7: 512–518.

Badr A and El-Shazly H (2012) Molecular approaches to origin, ancestry and domestication history of crop plants: Barley and clover as examples. J. Genet. Eng. Biotechnol. 10: 1–12.

Badr A, Müller K, Schäfer-Pregl R, Rabey HEl, Effgen S, Ibrahim HH, Pozzi C, Rohde W and Salamini F (2000) On the origin and domestication history of barley (*Hordeum vulgare*). Mol. Biol. Evol. 17: 499–510.

Bagavathiannan MV, Gulden RH, Begg GS and Van Acker RC (2010) The demography of feral alfalfa (*Medicago sativa* L.) populations occurring in roadside habitats in Southern Manitoba, Canada: implications for novel trait confinement. Environ. Sci. Pollut. Res. 17: 1448–1459.

Bagavathiannan MV, Gulden RH and Van Acker RC (2011) Occurrence of alfalfa (*Medicago sativa* L.) populations along roadsides in southern Manitoba, Canada and their potential role in intraspecific gene flow. Transgenic Res. 20: 397–407.

Bagavathiannan MV and Van Acker RC (2009) The biology and ecology of feral alfalfa (*Medicago sativa* L.) and its implications for novel trait confinement in North America. Critic. Rev. Pl. Sci. 28: 69–87.

Baik BK and Ullrich SE (2008) Barley for food: Characteristics, improvement, and renewed interest. J. Cereal. Sci. 48: 233–242.

Baltazar BM, Sanchez-Gonzalez JD, de la Cruz-Larios L and Schoper JB (2005) Pollination between maize and teosinte: an important determinant of gene flow in Mexico. Theor. Appl. Genet. 110: 519–526.

Bannert M, Vogler A and Stamp P (2008) Short-distance cross-pollination of maize in a small-field landscape as monitored by grain color markers. Europ. J. Agron. 29: 29–32.

Barnes DK, Goplen BP and Baylor JE (1988) Highlights in the USA and Canada. pp. 1–24. *In*: Hanson AA, Barnes DK and Hill Jr RR (eds.). Alfalfa and Alfalfa Improvement. Madison, WI: American Society of Agronomy Inc. Publishers.

Baum BR (1977) Oats: Wild and cultivated, a monograph of the genus *Avena* L. (Poaceae). Biosystematics Research Institute, Canada. Department of Agriculture, Monograph no. 14, Ottawa, ON, Canada.

Beckie HJ, Harker KN, Hall LM, Warwick SI, Legere A and Sikkema PH (2006) A decade of herbicide-resistant crops in Canada. Can. J. Plant Sci. 86: 1243–1264.

Berthaud J and Gepts P (2004) Assessment of effects on genetic diversity. CEC Secretariat Article 13 Report, Chapter 3 *In*: CEC (Commission for Environmental Cooperation) (ed.). Maize and Biodiversity: The Effects of Transgenic Maize in Mexico. Quebec, PQ, Canada: Communication Department of the CEC Secretariat.

Berville A, Breton C, Cunliffe K, Darmency H, Good AG, Gressel J, Hall LM, McPherson MA, Medail F, Pinatel C, Vaughan DA and Warwick SI (2005) Issues of ferality or potential for ferality in oats, olives, the pigeon-pea group, ryegrass species, safflower, and sugarcane. pp. 231–255. *In*: Gressel J (ed.). Crop Ferality and Volunteerism: A Threat to Food Security in the Transgenic Era. Boca Raton, FL: CRC Press.

Bervillé A, Muller MH, Poinso B and Serieys H (2004) Ferality: risks of gene flow between sunflower and other *Helianthus* species. pp. 209–229. *In*: Gressel J (ed.). Crop Ferality and Volunteerism: A Threat to Food Security in the Transgenic Era? CRC Press, Boca Raton, USA.

Bickelmann U and Leist N (1988) Homogeneity of oat cultivars with respect to outcrossing. pp. 358–363. *In*: Proceedings of the 3rd International Oat Conference. Lund, Sweden.

Bidinger FR, Mahalakshimi V and Rao GDP (1987) Assessment of drought resistance in pearl millet [*Pennisetum americanum* (L.) Leeke). II estimation of genotype response to stress. Aust. J. Agric. Res. 38: 49–59.

Billot C, Ramu P, Bouchet S, Chantereau J, Deu M, Gardes L, Noyer JL, Rami JF, Rivallan R, Li Y, Lu P, Wang T, Folkertsma RT, Arnaud E, Upadhyaya HD, Glaszmann JC and Hash CT (2013) Massive sorghum collection genotyped with SSR markers to enhance use of global genetic resources. PLoS One 8: 1–16.

Blackman BK, Scascitelli M, Kane NC, Luton HH, Rasmussen DA, Bye RA, Lentz DL and Rieseberg LH (2011) Sunflower domestication alleles support single domestication center in eastern North America. Proc. Natl. Acad. Sci. USA 108: 14360–14365.

Boerma HR and Moradshahi A (1975) Pollen movement within and between rows to male-sterile soybean. Crop Sci. 15: 858–861.

Bolton JL (1962) Alfalfa: Botany, Cultivation and Utilization. New York: Intel Science Publishers.

Bolton JL, Goplen BP and Baenziger H (1972) World distribution and historical developments. pp. 1–34. *In*: Hanson CH (ed.). Alfalfa Science and Technology. Madison, WI: American Society of Agronomy Inc. Publishers.

Brown AHD, Zohary D and Nevo E (1978) Outcrossing rates and heterozygosity in natural populations of *Hordeum spontaneum* Koch in Israel. Heredity 41: 49–62.

Brubaker CL and Wendel JF (1994) Reevaluating the origin of domesticated cotton (*Gossypium hirsutum*; Malvaceae) using nuclear restriction fragment length polymorphisms (RFLPs). Am. J. Bot. 81: 1309–1326.

Brunken JN (1977) A systematic study of *Pennisetum* sect. *Pennisetum* (Gramineae). Am. J. Bot. 64: 161–176.

Brunken J, de Wet JMJ and Harlan JR (1977) The morphology and domestication of pearl millet. Econ. Bot. 31: 163–174.

Burke JM, Gardner KA and Rieseberg LH (2002) The potential for gene flow between cultivated and wild sunflower (*Helianthus annuus*) in the United States. Am. J. Bot. 89: 1550–1552.

Burton GW (1974) Factors affecting pollen movement and natural crossing in pearl millet. Crop Sci. 14: 802–805.

Busconi M, Rossi D, Lorenzoni C, Baldi G and Fogher C (2012) Spread of herbicide-resistant weedy rice (red rice, *Oryza sativa* L.) after 5 years of Clearfield rice cultivation in Italy. Plant Biol. 14: 751–759.

Callaway E (2014) The birth of rice. Nature 514: 58–59.

Cantamutto M and Poverene M (2007) Genetically modified sunflower release: opportunities and risks. Field Crops Res. 101: 133–144.

Cao Q, Lu BR, Xia H, Rong J, Sala F, Spada A and Grassi F (2006) Genetic diversity and origin of weedy rice (*Oryza sativa* f. *spontanea*) populations found in North-eastern China revealed by simple sequence repeat (SSR) markers. Ann. Bot. 98: 1241–1252.

Chandler JM, Jan C and Beard BH (1986) Chromosomal differentiation among the annual *Helianthus* species. Syst. Bot. 11: 354–371.

Chang TT (1976) The origin, evolution, cultivation, dissemination, and diversification of Asian and African rices. Euphytica 25: 425–441.

Charmet G (2011). Wheat domestication: Lessons for the future. C. R. Biologies 334: 212–220.

Chen LJ, Lee DS, Song ZP, Suh HS and Lu BR (2004) Gene flow from cultivated rice (*Oryza sativa*) to its weedy and wild relatives. Ann. Bot. 93: 67–73.

Chen S, Nelson MN, Ghamkar K, Fu T and Cowling WA (2008). Divergent patterns of allelic diversity from similar origins: the case of oilseed rape (*Brassica napus* L.) in China and Australia. Genome 51: 1–10.

Chiang YC and Kiang YT (1987) Geometric position of genotypes, honeybee foraging patterns, and outcrossing in soybean. Bot. Bull. Acad. Sin. 28: 1–11.

Clayton WD and Renvoize SA (1982) Gramineae. pp. 451–898. *In:* Polhill RM (ed.). Flora of Tropical Africa, Part 3. Rotterdam, The Netherlands: AA Balkema on behalf of the East African Government.

Clotault J, Thuillet AC, Buiron M, De Mita S, Couderc M, Haussmann BIG, Mariac C and Vigouroux Y (2012) Evolutionary history of pearl millet (*Pennisetum glaucum* [L.] R. Br.) and selection on flowering genes since its domestication. Mol. Biol. Evol. 29: 1199–1212.

Coffman FA and Wiebe GA (1930) Unusual crossing in oats in Aberdeen, Idaho. J. Amer. Soc. Agron. 22: 245–250.

Conway G (1997) The Doubly Green Revolution: Food for All in the 21st Century. Ithaca, NY: Cornell University Press. pp. 31–43.

Cotton Incorporated (2015) World Cotton Production. http://www.cottoninc.com/corporate/MarketData/MonthlyEconomicLetter/pdfs/English-pdf-charts-and-tables/World-Cotton-Production-Bales.pdf. Accessed November 10, 2015.

Cresswell JE (1999) The influence of nectar and pollen availability on pollen transfer by individual flowers of oilseed rape (*Brassica napus*) when pollinated by bumble bees. J. Ecol. 87: 670–677.

Dai F, Nevo E, Wu D, Comadran J, Zhou M, Qiu L, Chen Z, Beiles A, Chen G and Zhang G (2012) Tibet is one of the centers of domestication of cultivated barley. Proc. Natl. Acad. Sci. USA 109: 16969–16973.

Devos Y, Schrijver AD and Reheul D (2009). Quantifying the introgressive hybridization propensity between transgenic oilseed rape and its wild/weedy relatives. Environ. Monit. Assess. 149: 303–322.

de Wet JMJ (1978) Systematics and evolution of *Sorghum* sect. *Sorghum* (Gramineae). Am. J. Bot. 65: 477–484.

de Wet JMJ, Brink DE and Cohen CE (1983) Systematics of *Tripsacum* section Fasciculata (Gramineae). Amer. J. Bot. 70: 1139–1146.
de Wet JMJ and Harlan JR (1971) The origins and domestication of *Sorghum bicolor*. Econ. Bot. 25: 128–135.
de Wet JMJ, Harlan JR and Brink DE (1982) Systematics of *Tripsacum dactyloides* (Gramineae). Amer. J. Bot. 69: 1251–1257.
de Wet JMJ, Harlan JR, Engle LM and Grant CA (1973) Breeding behavior of maize-*Tripsacum* hybrids. Crop Sci. 13: 254–256.
Dedio W and Putt ED (1980) Sunflower. pp. 631–644. *In*: Fehr WR and Hadley HH (eds.). Hybridization in Crop Plants. Madison, WI: American Society of Agronomy Inc. Publishers.
Defelice MS (2006) Shattercane, *Sorghum bicolor* (L.) Moench ssp. *drummondii* (Nees ex Steud.) de Wet ex Davidse–Black sheep of the family. Weed Technol. 20: 1076–1083.
Diamond J (1999) Guns, Germs and Steel: The Fates of Human Societies. New York: W.W. Norton & Company. pp. 93–103.
Dicko MH, Gruppen H, Traoré AS, Voragen AGJ and Berkel WJH Van (2006) Sorghum grain as human food in Africa : Relevance of content of starch and amylase activities. African J. Biotechnol. 5: 384–395.
Doebley JF (1990) Molecular evidence and the evolution of maize. Econ. Bot. 44: 6–27.
Doebley FJ and Iltis HH (1980) Taxonomy of *Zea* (Gramineae): I. A subgeneric classification with key to taxa. Amer. J. Bot. 67: 982–993.
Doebley J, von Bothmer R and Larson S (1992) Chloroplast DNA variation and the phylogeny of *Hordeum* (Poaceae). Am. J. Bot. 79: 576–584.
Dorado O, Rieseberg LH and Arias DM (1992) Chloroplast DNA introgression in Southern California sunflowers. Evolution 46: 566–572.
Dorokhov D, Ignatov A, Deineko E, Serjapin A, Ala A and Skryabin K (2004) The chance for gene flow from herbicide-resistant GM soybean to wild soy in its natural inhabitation at Russian far east region. pp. 151–161. *In*: den Nijs HCM, Bartsch D and Sweet J (eds.). Introgression from Genetically Modified Plants into Wild Relatives. Wallingford, UK: CAB International.
Dowling LF, Arndt C and Hamaker BR (2002) Economic viability of high digestibility sorghum as feed for market broilers. Agron. J. 94: 1050–1058.
Duke JA (1981) Handbook of Legumes of World Economic Importance. New York: Plenum Press. 345 p.
Duke SO (2005) Taking stock of herbicide-resistant crops ten years after introduction. Pest Manag. Sci. 61: 211–218.
Dvorak J, di Terlizzi P, Zhang HB and Resta P (1993) The evolution of polyploidy wheats: Identification of the A genome donor species. Genome 36: 21–31.
Dweikat I (2005) A diploid, interspecific, fertile hybrid from cultivated sorghum, *Sorghum bicolor*, and the common Johnsongrass weed *Sorghum halepense*. Mol. Breeding 16: 93–101.
Eastham K and Sweet J (2002) Genetically modified organisms (GMOs): The significance of gene flow through pollen transfer. Copenhagen, Denmark: European Environment Agency. 9 p.
Ejeta G and Grenier C (2005) Sorghum and its weedy hybrids. pp. 123–135. *In*: Gressel JB (ed.). Crop Ferality and Volunteerism. Boca Raton, FL: CRC Press.
Ellstrand NC and KA Schierenbeck (2006) Hybridization as a stimulus for the evolution of invasiveness in plants. Euphytica 148: 35–46.
Ellstrand NC (2003) Dangerous liaisons? When cultivated plants mate with their wild relatives. The Johns Hopkins University Press, Baltimore, Maryland, 244 p.
Evenson RE and Gollin D (2003). Assessing the impact of the Green Revolution 1960 to 2000. Science 300: 758–762.
[FAOSTAT] Food and Agriculture Organization of the United Nations Statistics Division (2013) Crop production data. http://faostat3.fao.org/home/E. Accessed October 8, 2015.
Faure N, Serieys H and Berville A (2002) Potential gene flow from cultivated sunflower to volunteer, wild *Helianthus* species in Europe. Agric. Ecosyst. Environ. 89: 183–190.

Feleke Y, Pasquet RS and Gepts P (2006) Development of PCR-based chloroplast DNA markers that characterize domesticated cowpea (*Vigna unguiculata* subsp. *unguiculata* var. *unguiculata*) and highlight its crop-weed complex. Plant Syst. Evol. 262: 75–87.

FitzJohn RG, Armstrong TT, Newstrom-Lloyd LE, Wilton AD and Cochrane M (2007) Hybridization within *Brassica* and allied genera: evaluation of potential for transgene escape. Euphytica 158: 209–230.

Fryxell PA (1979) The natural history of the cotton tribe (Malvaceae, Tribe Gossypieae). College Station, TX: Texas A&M University Press.

Fujita R, Ohara M, Okazaki K and Shimamoto Y (1997) The extent of natural pollination in wild soybean (Glycine soja). J. Hered. 76: 373–374.

Fuller DQ, Harvey E and Qin L (2007) Presumed domestication? Evidence for wild rice cultivation and domestication in the fifth millennium BC of the lower Yangtze region. Antiquity 81: 316–331.

Gabard JM and Huby JP (2001) Sunfonylurea-tolerant sunflower plants. United States Patent Application # 20050044587-2004.

Gatford KT, Basri Z, Edlington J, Lloyd J, Qureshi JA, Brettell R and Fincher GB (2006) Gene flow from transgenic wheat and barley under field conditions. Euphytica 151: 383–391.

Ge S, Sang T, Lu BR and Hong DY (1999) Phylogeny of rice genomes with emphasis on origins of allotetraploid species. Proc. Natl. Acad. Sci. USA 96: 14400–14405.

Gealy DR, Mitten DH and Rutger JN (2003) Gene flow between red rice (*Oryza sativa*) and herbicide-resistant rice (*O. sativa*): Implications for weed management. Weed Technol. 17: 627–645.

GMO Compass (2010) GMO plant database. http://www.gmo-compass.org/eng/database/plants/36.barley.html. Accessed October 8, 2015.

Gotmare V, Singh P and Tule BN (2015) Wild and cultivated species of cotton. Central Institute for Cotton Research, Nagpur, India. Technical Bulletin No. 5. http://www.cicr.org.in/pdf/wild_species%20.pdf. Accessed November 10, 2015.

Goulart ICGR, Borba TCO, Menezes VG and Merotto Jr A (2014) Distribution of weedy red rice (*Oryza sativa*) resistant to imidazolinone herbicides and its relationship to rice cultivars and wild *Oryza* Species. Weed Sci. 62: 280–293.

Govindaraj M, Rai KN, Shanmugasundaram P, Dwivedi SL, Sahrawat KL, Muthaiah AR and Rao AS (2013) Combining ability and heterosis for grain iron and zinc densities in pearl millet. Crop Sci. 53: 507–517.

Groves R, Hoskings JR, Batianoff GN, Cooke DA, Cowie ID, Keighery GJ, Lepshci BJ, Rozefelds AC and Walsh NG (2000) The naturalised non-native flora of Australia: its categorisation and threat to native plant biodiversity, Unpublished report to Environment Australia by the CRC for Weed Management Systems.

Gupta SC (1999) Seed Production Procedures in Sorghum and Pearl Millet. Patancheru, India: International Crops Research Institute for the Semi-Arid Tropics, Information Bulletin No. 58: 12 p.

Gupta SK and Pratap A (2007) History, origin and evolution. Adv. in Bot. Res. 45: 1–20.

Hammer GL and Muchow RC (1994) Assessing climatic risk to sorghum production in water-limited subtropical environments I. Development and testing of a simulation model. Field Crop Res. 36: 221–234.

Hanson AA, Barnes DK and Hill Jr RR (1988) Alfalfa and Alfalfa Improvement. Madison, WI: American Society of Agronomy Inc. Publishers.

Hanson BD, Mallory-Smith CA and Price WJ (2005) Interspecific hybridization: Potential for movement of herbicide resistance from wheat to jointed goatgrass (*Aegilops cylindrica*) Weed Technol. 19: 674–682.

Harlan JR (1971) Agricultural origins: Centers and noncenters. Science 174: 468–474.

Harlan JR (1977) The origins of cereal agriculture in the old World. pp. 357–383. *In*: Reed CA (ed.). Origins of Agriculture. The Hague, Netherlands: Moulton Publishing Co.

Harlan JR and de Wet JMJ (1971) Toward a rational classification of cultivated plants. Taxon. 20: 509–517.

Harlan JR, de Wet JMJ and Stemler ABL (1976) Plant domestication and indigenous African agriculture. pp. 1–9. *In*: Harlan JR, de Wet JMJ and Stemler ABL (eds.). Origins of African Domestication. The Hague: Mouton Publishers.

Harlan JR and Zohary D (1966) Distribution of wild wheats and barley—The present distribution of wild forms may provide clues to the regions of early cereal domestication. Science 153: 1074–1080.

Heiser CB, Smith DM, Clevenger S and Martin WC (1969) The North American sunflowers (*Helianthus*). Mem. Torrey Bot. Club 22: 1–218.

Holm LG, Plucknett DL, Pancho JV and Herberger JP (1977) The World's Worst Weeds: Distribution and Biology. Honolulu, HI: University Press of Hawaii. 609 p.

Hutchinson JB (1951) Intra-specific differentiation in *Gossypium hirsutum*. Heredity 5: 161–193.

Hutchinson JB (1959) The Application of Genetics to Cotton Improvement. Cambridge: Cambridge University Press.

Hymowitz T, Singh RJ and Kollipara KP (1998) The genomes of the Glycine. Plant Breed Rev. 16: 289–317.

Hymowitz T (1990) Soybeans: The success story. pp. 159–163. *In*: Janick J and Simon JE (eds.). Advances in New Crops. Portland: Timber Press.

Hymowitz T and Harlan JR (1983) Introduction of soybean to North America by Samuel Bowen in 1765. Econ. Bot. 37: 371–379.

[IRGSP] International Rice Genome Sequencing Project (2005) IRGSP Releases the Assembled Rice Genome Sequences. http://rgp.dna.affrc.go.jp/IRGSP/Build2/build2.html. Accessed October 31, 2015.

[ISAAA] International Service for the Acquisition of Agri-Biotech Applications (2015) GM Approval Database: Rice. http://www.isaaa.org/gmapprovaldatabase/advsearch/default.asp?CropID=17&TraitTypeID=Any&DeveloperID=Any&CountryID=Any&ApprovalTypeID=Any. Accessed October 31, 2015.

Jakob SS, Rödder D, Engler JO, Shaaf S, Özkan H, Blattner FR and Kilian B (2014) Evolutionary history of wild barley (*Hordeum vulgare* subsp. *spontaneum*) analyzed using multilocus sequence data and paleodistribution modeling. Genome Biol. Evol. 6: 685–702.

Jhala AJ and LM Hall (2013). Risk assessment of herbicide resistant crops with special reference to pollen mediated gene flow. pp. 237–254. *In*: Price AJ and Kelton JA (eds.). Herbicides—Advances in Research. In Tech Scientific Publisher, 41 Madison Avenue, Manhattan, New York.

Jhala AJ, Bhatt H, Keith A and Hall LM (2011) Pollen mediated gene flow in flax: Can genetically engineered and organic flax coexist? Heredity 106: 557–56.

Jhala AJ, Hall LM and Hall JC (2008) Potential hybridization of flax (*Linum usitatissimum* L.) with weedy and wild relatives: An avenue for movement of engineered genes? Crop Science 48: 825–840.

Johnson J, Haley S and Westra P (2002). Production: Clearfield wheat. Crop Series: Colorado State University Cooperative Extension 3: 116.

Kantar MB, Sosa CC, Khoury CK, Castaneda-Alvarez NP, Achicanoy HA, Bernau V, Kane NC, Marek L, Seiler G and Rieseberg LH (2015) Ecogeography and utility to plant breeding of the crop wild relatives of sunflower (*Helianthus annuus* L.). Front Plant Sci. (in press) http://dx.doi.org/10.3389/fpls.2015.00841.

Kareiva P, Morris W and Jacobi CM (1994) Studying and managing the risk of cross-fertilization between transgenic crops and wild relatives. Mol. Ecol. 3: 15–21.

Kermicle JL and Allen JO (1990) Cross-incompatibility between maize and teosinte. Maydica 35: 399–408.

Kermicle JL (1997) Cross compatibility within the genus *Zea*. pp. 40–43. *In*: Serratos JA, Willcox MC and Castillo F (eds.). Proceedings of a Forum: Gene Flow Among Maize Landraces, Improved Maize Varieties, and Teosintes; Implications for Transgenic Maize. Mexico, DF: International Maize and Wheat Improvement Center [CIMMYT].

Kholová J, Hash CT, Kakkera A, Kočová M and Vadez V (2010) Constitutive water-conserving mechanisms are correlated with the terminal drought tolerance of pearl millet [*Pennisetum glaucum* (L.) R. Br.]. J. Exp. Bot. 61: 369–377.

Khush GS (1997) Origin, dispersal, cultivation and variation of rice. Plant Mol. Biol. 35: 25–34.

Kiang YT, Chiang YC and Kaizuma N (1992) Genetic diversity in natural populations of wild soybean in Iwate Prefecture, Japan. J. Hered. 83: 325–329.

Kim S and Dale BE (2004) Global potential bioethanol production from wasted crops and crop residues. Biomass and Bioenergy 26: 361–375.

Kimber CT (2000) Origins of domesticated sorghum and its early diffusion to India and China. pp. 3–98. *In*: Smith CW and Frederiksen RA (eds.). Sorghum: Origin, History, Technology, and Production. New York: John Wiley & Sons.

Kimber G and Feldman M (1987). Wild wheat: An introduction. Special report 353. Columbia, MO: College of Agriculture, University of Missouri-Columbia. 142 p.

Klinkowski M (1933) Lucerne: its ecological position and distribution in the world. Bull. Imp. Bur. Pl. Genet. Herb. Pl. 12: 61 p.

Knispel AL, McLachlan SL, Van Acker RC and Friesen LF (2008) Gene flow and multiple herbicide resistance in escaped canola populations. Weed Sci. 56: 72–80.

Komatsuda T, Tanno KI, Salomon B, Bryngelsson T and von Bothmer R (1999) Phylogeny in the genus *Hordeum* based on nucleotide sequences closely linked to the vrs1 locus (row number of spikelets). Genome 42: 973–981.

Ladizinsky G (1988) A new species of oat from Sicily, possibly the tetraploid progenitor of hexaploid oats. Genet. Resour. Crop Evol. 45: 263–269.

Ladizinsky G and Fainstein R (1977) Introgression between the cultivated hexaploid oat *Avana sativa* and the tetraploid wild oats *A. magna* and *A. murphyi*. Can. J. Genet. Cytol. 19: 59–66.

Ladizinsky G and Zohary D (1971) Notes on species delimitation, species relationships, and polyploidy in *Avena*. Euphytica 20: 380–395.

Langevin SA, Clay K and Grace JB (1990) The incidence and effects of hybridization between cultivated rice and its related weed red rice (*Oryza sativa* L.). Evolution 44: 1000–1008.

Legget JM and Thomas H (1995) Oat evolution and cytogenetics. pp. 120–149. *In*: Welch RW (ed.). The Oat Crop: Production and Utilization. London: Chapman and Hall.

Lentz D, Pohl M, Alvarado JL, Tarigaht S and Bye R (2008) Sunflower (*Helianthus annuus* L.) as Pre-Columbian Domesticate in Mexico. Proc. Natl. Acad. Sci. USA 107: 6232–6237.

Lentz D, Pohl M, Pope K and Wyatt A (2001) Prehistoric sunflower (*Helianthus annuus* L.) domestication in Mexico. Econ. Bot. 55: 370–376.

Leuck DB and Burton GW (1966) Pollination of pearl millet by insects. J. Econ. Entomol. 59: 1308–1309.

Linder CR, Taha I, Seiler GJ, Snow AA and Rieseberg LH (1998) Long-term introgression of crop genes into wild sunflower populations. Theor. Appl. Genet. 96: 339–347.

Lu BR and Snow AA (2005) Gene flow from genetically modified rice and its environmental consequences. BioSci. 55: 669–678.

Ma B, Subedi K and Reid L (2004) Extent of cross-pollination in maize by pollen from neighboring transgenic hybrids. Crop Sci. 44: 1273–1282.

Ma XY, Li C, Wang AD, Duan RJ, Jiao GL, Nevo E and Chen GX (2012) Genetic diversity of wild barley (*Hordeum vulgare* ssp. *spontaneum*) and its utilization for barley improvement. Sci. Cold Arid Reg. 4: 453–461.

Mac Key J (2005) Wheat: Its concept, evolution, and taxonomy. pp. 3–61. *In*: Royo C, Nachit MM, Di Fonzo N, Araus JL, Pfeiffer WH and Slafer GA (eds.). Durum Wheat Breeding: Current Approaches and Future Strategies, Vol. I, Binghamton, NY: Food Products Press.

Mallory-Smith C and Zapiola M (2008) Gene flow from glyphosate-resistant crops. Pest Manag. Sci. 64: 428–440.

Manning K, Pelling R, Higham T, Schwenniger JL and Fuller DQ (2011) 4500-Year old domesticated pearl millet (*Pennisetum glaucum*) from the Tilemsi Valley, Mali: New insights into an alternative cereal domestication pathway. J. Archaeol. Sci. 38: 312–322.

Mariac C, Robert T, Allinne C, Remigereau MS, Luxereau A, Tidjani M, Seyni O, Bezancon G, Pham JL and Sarr A (2006) Genetic diversity and gene flow among pearl millet crop/weed complex: a case study. Theor. Appl. Genet. 113: 1003–1014.

Massinga RA, Al-Khatib K, St. Amand P and Miller JF (2003) Gene flow from imidazolinone-resistant domesticated sunflower to wild relatives. Weed Sci. 51: 854–862.

Matsuoka Y, Vigouroux Y, Goodman MM, Sanchez J, Buckler ES and Doebley JF (2002) A single domestication for maize shown by multilocus microsatellite genotyping. Proc. Natl. Acad. Sci. USA 99: 6080–84.

Matus-Cadiz MA, Hucl P and Dupuis B (2007). Pollen-mediated gene flow in wheat at the commercial scale. Crop Sci. 47: 573–581.

McFadden ES and Sears ER (1946). The origin of *Triticum spleta* and its free threshing hexaploid relatives. J. of Hered. 37: 81–116.

McGregor SE (1976) Insect Pollination of Cultivated Crop Plants. United States Department of Agriculture, Agricultural Research Service, Agriculture Handbook 496. Washington, DC: Government Printing Office 411 p.

Mendelsohn M, Kough J, Vaituzis V and Matthews K (2003) Are *Bt* crops safe? Nat. Biotech. 21: 1003–1009.

Mesquida J, Renard M and Pierre JS (1982) Rapeseed (*Brassica napus* L.) productivity: The effect of honeybees and different pollination conditions in cage and field tests. Apidol. 19: 51–72.

Messeguer J, Fogher C, Guiderdoni E, Marfà V, Català MM, Baldi G and Melé E (2001) Field assessments of gene flow from transgenic to cultivated rice (*Oryza sativa* L.) using a herbicide resistance gene as tracer marker. Theor. Appl. Genet. 103: 1151–1159.

Miller JF and Al-Khatib K (2004) Registration of two oilseed sunflower genetic stocks, SURES-1 and SURES-2, resistant to tribenuron herbicide. Crop Sci. 44: 1037–1038.

Moffett JO, Stith LS, Burkhart CC and Shipman CW (1975) Honey bee visits to cotton flowers. Environ. Entomol. 4: 203–206.

Molina-Cano JL, Fra-Mon P, Salcedo G, Aragoncillo C, Togores FR de and García-Olmedo F (1987) Morocco as a possible domestication center for barley: biochemical and agromorphological evidence. Theor. Appl. Genet. 73: 531–536.

Mrízová K, Holasková E, Öz MT, Jiskrová E, Frébort I and Galuszka P (2014) Transgenic barley: A prospective tool for biotechnology and agriculture. Biotechnol. Adv. 32: 137–157.

Msongaleli B, Rwehumbiza F, Tumbo SD and Kihupi N (2014) Sorghum yield response to changing climatic conditions in semi-arid central Tanzania: Evaluating crop simulation model applicability. Agric. Sci. 5: 822–833.

Nagaharu U (1935) Genome analysis in *Brassica* with special reference to the experimental formation of *B. napus* and peculiar mode of fertilization. Japan J. Bot. 7: 389–452.

NAAIC (2000) Importance of alfalfa. North American Alfalfa Improvement Conference. https://www.naaic.org/Alfalfa/Importance.html. Accessed November 22, 2015.

Newton AC, Flavell AJ, George TS, Leat P, Mullholland B, Ramsay L, Revoredo-Giha C, Russell J, Steffenson BJ, Swanston JS, Thomas WTB, Waugh R, White PJ and Bingham IJ (2011) Crops that feed the world 4. Barley: a resilient crop? Strengths and weaknesses in the context of food security. Food Secur. 3: 141–178.

Normile D (2004) Yangtze seen as earliest rice site. Science 275: 309.

NSA (2015) Sunflower Statistics: World Supply & Disappearance. National Sunflower Association. http://www.sunflowernsa.com/stats/world-supply. Accessed November 7, 2015.

Nuijten E and Richards P (2011) Pollen flows within and between rice and millet fields in relation to farmer variety development in the Gambia. Pl. Gen. Res. 9: 361–374.

Oka HI (1974) Experimental studies on the origin of cultivated rice. Genetics 78: 475–486.

Oka HI (1983) Genetic control of regenerating success in semi-natural conditions observed among lines derived from a cultivated x wild soybean hybrid. J. Appl. Ecol. 20: 937–949.

Oumar I, Mariac C, Pham JL and Vigouroux Y (2008) Phylogeny and origin of pearl millet (*Pennisetum glaucum* [L.] R. Br) as revealed by microsatellite loci. Theor. Appl. Genet. 117: 489–497.

Palmer RG, Gai J, Sun H and Burton JW (2001) Production and evaluation of hybrid soybean. Plant Breed. Rev. 21: 263–307.

Panella L and Gepts P (1992) Genetic relationship within *Vigna unguiculata* (L.) Walp. based on isozyme analysis. Genet. Resour. Crop Evol. 39: 71–88.

Pantulu JV (1967) Pachytene pairing and meiosis in the F1 hybrid of *Pennisetum typhoides* and *P. purpureum.* Cytologia 32: 532–541.

Parzies HK, Spoor W and Ennos RA (2000) Outcrossing rates of barley landraces from Syria. Pl. Breed 119: 520–522.

Pasquet RS (1993a) Classification infraspe´cifique des formes spontane´es de *Vigna unguiculata* (L.) Walp. a` partir de donne´es morphologiques. Bull. Jard. Bot. Nat. Belg. 62: 127–173.

Pasquet RS (1993b) Two new subspecies of *Vigna unguiculata* (L.) Walp. (*Leguminosae: Papilionoideae*). Kew Bull. 48: 805–806.

Pasquet RS (1997) A new subspecies of *Vigna unguiculata* (*Leguminosae-Papilionoideae*). Kew Bull. 52: 840.

Pasquet RS (1999) Genetic relationships among subspecies of *Vigna unguiculata* (L.) Walp. based on allozyme variation. Theor. Appl. Genet. 98: 1104–1119.

Paterson AH, Schertz KF, Lin YR, Liu SC and Chang YL (1995) The weediness of wild plants: Molecular analysis of genes influencing dispersal and persistence of johnsongrass, *Sorghum halepense* (L.) Pers. Proc. Natl. Acad. Sci. USA 92: 6127–6131.

Pedersen JF, Toy JJ and Johnson B (1998) Natural outcrossing of sorghum and sudangrass in the central Great Plains. Crop Sci. 38: 937–939.

Poehlmann JM (1959) Breeding of Field Crop. New York: Henry Holt. 427 p.

Potrykus I (2001) The 'Golden Rice' tale. *In Vitro* Cellular and Developmental Biology—Plant 37: 93–100.

Poverene M, Carrera A, Ureta S and Cantamutto M (2004) Wild *Helianthus* species and wild-sunflower hybridization in Argentina. Helia 27: 133–142.

Putnam DH (2006) Methods to enable coexistence of diverse production systems involving genetically engineered alfalfa. University of California Agriculture and Natural Resources Publication No. 8193: 9 p.

Quiros CF and Bauchan GR (1988) The genus *Medicago* and the origin of the Medicago sativa complex. pp. 93–124. *In:* Hanson AA, Barnes DK and Hill Jr RR (eds.). Alfalfa and Alfalfa Improvement. Madison, WI: American Society of Agronomy Inc. Publishers.

Rao GM, Nadre KR and Suryanarayana MC (1996) Studies on the utility of honey bees on production of foundation seed of cotton cv. NCMHH-20. Indian Bee J. 58: 13–15.

Raymer PL (2002) Canola: An emerging oilseed crop. pp. 122–126. *In:* Janick J and Whipkey A (eds.). Trends in New Crops and New Uses. Alexandria, VA: ASHS Press.

Raynor GS, Ogden EC and Hayes JV (1972) Dispersion and deposition of corn pollen from experimental sources. Agron. J. 64: 420–427.

Redden RJ (1977). Natural outcrossing in wheat substation lines. Aust. J. of Ag. Res. 28: 763–768.

Ren X, Nevo E, Sun D and Sun G (2013) Tibet as a potential domestication center of cultivated barley of China. PLoS One 8: 1–7.

Richards JS, Stanley JN and Gregg PC (2005) Viability of cotton and canola pollen on the proboscis of *Helicoverpa armigera*: implications for spread of transgenes and pollination ecology. Ecol. Entomol. 30: 327–333.

Richmond TR (1951) Procedures and methods of cotton breeding with special reference to American cultivated species. Adv. Genet. 4: 213–245.

Riley R, Unrau J and Chapman V (1958). Evidence of the origin of the B genome of wheat. J. of Hered. 49: 91–98.

Ritala A, Nuutila AM, Aikasalo R, Kauppinen V and Tammisola J (2002) Measuring gene flow in the cultivation of transgenic barley. Crop Sci. 42: 278–285.

Robert T, Lamy F and Sarr A (1992) Evolutionary role of gametophytic selection in the domestication of *Pennisetum typhoides* (pearl millet): A two-locus asymmetric model. Heredity 69: 372–381.

Rong J, Xia H, Zhu YY, Wang YY and Lu BR (2004) Asymmetric gene flow between traditional and hybrid rice varieties (*Oryza sativa*) estimated by nuclear SSRs and its implication in germplasm conservation. New Phytol. 163: 439–445.

Sala CA, Bulos M, Altieri E and Ramos ML (2012) Genetics and breeding of herbicide tolerance in sunflower. Helia 35: 57–50.

Sarr A, Sandmeier M and Pernès J (1988) Gametophytic competition in pearl millet, *Pennisetum typhoides* (Burm.) Stapf & Hubb. Genome 30: 924–929.

Sasaki T, Matsumoto T, Yamamoto K, Sakata K, Baba T, Katayose Y, Wu J, Niimura Y, Cheng Z, Nagamura Y, Antonio Ba, Kanamori H, Hosokawa S, Masukawa M, Arikawa K, Chiden Y, Hayashi M, Okamoto M, Ando T, Aoki H, Arita K, Hamada M, Harada C, Hijishita S, Honda M, Ichikawa Y, Idonuma A, Iijima M, Ikeda M, Ikeno M, Ito S, Ito T, Ito Y, Ito Y, Iwabuchi A, Kamiya K, Karasawa W, Katagiri S, Kikuta A, Kobayashi N, Kono I, Machita K, Maehara T, Mizuno H, Mizubayashi T, Mukai Y, Nagasaki H, Nakashima M, Nakama Y, Nakamichi Y, Nakamura M, Namiki N, Negishi M, Ohta I, Ono N, Saji S, Sakai K, Shibata M, Shimokawa T, Shomura A, Song J, Takazaki Y, Terasawa K, Tsuji K, Waki K, Yamagata H, Yamane H, Yoshiki S, Yoshihara R, Yukawa K, Zhong H, Iwama H, Endo T, Ito H, Hahn JH, Kim HI, Eun MY, Yano M, Jiang J and Gojobori T (2002) The genome sequence and structure of rice chromosome 1. Nature 420: 312–316.

Sato YI (2000) Origin and evolution of wild, weedy, and cultivated rice. pp. 7–15. *In*: Baki BB, Chin DV and Mortimer M. (eds.). Wild and Weedy Rice in Rice Ecosystems in Asia—A Review. Limited Proceedings No. 2. Los Baños, Philippines: International Rice Research Institute.

Schilling EE (2006) *Helianthus*. pp. 141–169. *In*: Flora of North America Editorial Committee (eds.). Flora of North America Committee, Vol. 21. New York: Oxford University Press.

Schilling EE and Heiser CB (1981) Intrgeneric classification of Helianthus (Compositae). Taxon. 30: 393–403.

Schmidt JJ, Pedersen JF, Bernards ML and Lindquist JL (2013) Rate of shattercane × sorghum hybridization *in situ*. Crop Sci. 53: 1677–1685.

Schober TJ, Messerschmidt M, Bean SR, Park SH and Arendt EK (2005) Gluten-free bread from sorghum: Quality differences among hybrids. Cereal Chem. 82: 394–404.

Second G (1982) Origin of the genic diversity of cultivated rice (*Oryza* spp.): Study of the polymorphism scored at 40 isozyme loci. Japanese J. Genet. 57: 25–57.

Singh RJ and Hymowitz T (1989) The genomic relationships between *Glycine soja* Sieb. and Zucc., *G. glycine* (L.) Merr., and *'G. gracilis'* Skvortz. Plant Breed 103: 171–173.

Skvortzow BW (1927) The soybean–wild and cultivated in Eastern Asia. Proceedings of Manchurian Research Society Publication Series A. Nat. Hist. Sec. 22: 1–8.

Smartt J (1990) The old world pulses. pp. 140–175. *In*: Smartt J (ed.). Grain Legumes: Evolution and Genetic Resources. Cambridge: Cambridge University Press.

Smith CW and Cothren JT (1999) Cotton: Origin, History, Technology, and Production. New York: John Wiley & Sons, Inc.

Snow AA, Rieseberg LH, Alexander HM, Cummings C and Pilson D (1998) Assessment of gene flow and potential effects of genetically engineered sunflowers on wild relatives. pp. 19–25. *In*: The Biosafety Results of Field Tests of Genetically Modified Plants and Microorganisms—5th Int. Symposium, Braunschweig, Germany.

Snow AA, Andersen B and Jorgensen RB (1999) Costs of transgenic herbicide resistance introgressed from *brassica napus* into weedy *B. rapa*. Molecular Ecology 8: 605–615.

Song ZP, Lu BR, Wang B and Chen JK (2004) Fitness estimation through performance comparison of F1 hybrids with their parental species *Oryza rufipogon* and *O. sativa*. Ann. Bot. 93: 311–316.

Song ZP, Lu BR, Zhu YG and Chen JK (2003) Gene flow from cultivated rice to the wild species *Oryza rufipogon* under experimental field conditions. New Phytol. 157: 657–665.

SoyStats (2015) A Reference Guide to Important Soybean Facts & Figures. A publication of the American Soybean Association. http://soystats.com/international-world-soybean-production. Accessed November 2, 2015.

St. Amand PC, Skinner DZ and Peaden RN (2000) Risk of alfalfa transgene dissemination and scale-dependent effects. Theor. Appl. Genet. 101: 107–114.

Stapf O and Hubbard CE (1934) Pennisetum. pp. 954–1070. *In*: Prain D (ed.). Flora of Tropical Africa, Part 9. London: Crown Agents.

Streit L (2012) DuPont™ ExpressSun® herbicide technology in sunflower. pp. 143–149. *In*: Proceedings of the 18th Sunflower Conference, Mar del Plata-Balcarce, Argentina.

Suh HS, Back J and Ha J (1997) Weedy rice occurrence and position in transplanted and direct-seeded farmer's fields. Korean J. Crop Sci. 42: 352–356.

Sullivan P, Arendt E and Gallagher E (2013) The increasing use of barley and barley by-products in the production of healthier baked goods. Trends Food Sci. Technol. 29: 124–134.

Tako E, Reed SM, Budiman J, Hart JJ and Glahn RP (2015) Higher iron pearl millet (*Pennisetum glaucum* L.) provides more absorbable iron that is limited by increased polyphenolic content. Nutr. J. 14: 11.

Techio VH, Davide LC and Pereira AV (2006) Meiosis in elephant grass (*Pennisetum purpureum*), pearl millet (*Pennisetum glaucum*) (Poaceae, Poales), and their interspecific hybrids. Genet. Mol. Biol. 29: 353–362.

Teuber L, Mueller S, Van Deynze AV, Fitzpatrick S, Hagler J and Arias J (2007) Seed-to-seed and hay-to-seed pollen mediated gene flow in alfalfa. Proceedings of the North Central Weed Science Society, St. Louis, MO.

Teuber LR, Van Deynze A, Mueller S, McCaslin M, Fitzpatrick S and Rogan G (2004) Gene flow in alfalfa under honeybee (*Apis mellifera*) pollination. Joint Meeting of the 39th North American Alfalfa Improvement Conference and 18th Trifolium Conference, Quebec City, Canada.

Thomas Jefferson Agricultural Institute [TJAI] (2010) Cowpea: a versatile legume for hot, dry conditions. Columbia, MO. https://hort.purdue.edu/newcrop/articles/ji-cowpea.html Accessed November 2, 2015.

Thomas H (1995) Oats. pp. 132–137. *In*: Simmonds S and Smartt J (eds.). Evolution of Crop Plants, Harlow, UK: Longman Scientific and Technical.

Ting YC (1963) A preliminary report on the 4th chromosome male gametophyte factor in teosintes. Maize Genet. Coop Newsletter 37: 6–7.

Tostain S (1992) Enzyme diversity in pearl millet (*Pennisetum glaucum* L.): 3. Wild millet. Theor. Appl. Genet. 83: 733–742.

Tuinstra MR, Al-Khatib K (inventors and assignee) (2010) May 6. Acetyl-CoA carboxylase herbicide resistant sorghum. US patent 20100115663 A1.

Tuinstra MR, Al-Khatib K (inventors); Kansas State University Research Foundation (assignee) (2008) September 4. Acetolactate synthase herbicide resistant sorghum. US patent 20080216187 A1.

USDA (2015) Plants Database. United States Department of Agriculture. http://plants.usda.gov. Accessed November 12, 2015.

Vadez V, Hash T, Bidinger FR and Kholova J (2012) II.1.5 Phenotyping pearl millet for adaptation to drought. Front Physiol. 3: 1–12.

Van Deynze A, Fitzpatrick S, Hammon B, McCaslin MH, Putnam DH, Teuber LR and Undersander DJ (2008) Gene flow in alfalfa: biology, mitigation, and potential impact on production. The Council for Agricultural Science and Technology, Ames, IA.

Van Deynze AE, Sundstrom FJ and Bradford KJ (2005) Pollen-mediated gene flow in California cotton depends on pollinator activity. Crop Sci. 45: 1565–1570.

Vara Prasad PV and Staggenborg SA (2010) Growth and production of sorghum and millets. pp. 1–34. *In*: Verheye WH (ed.). Soils, Plant Growth and Crop Production. Vol. II. Paris, France: Eolss Publishers.

Vavilov NI (1926) Studies on the origin of cultivated plants. Bull. Appl. Bot. Genet. Plant Breed USSR 16: 1–248.

Vavilov NI (1951) The origin, variation, immunity, and breeding of cultivated plants. Cheron. Bot. 13: 1–366.

Vavilov NI (1992) Origin and geography of cultivated plants. *In*: Dorofeyev VF (ed.). Translated by Love D, Cambridge: Cambridge University Press. 332 p.

Viands DR, Sun P and Barnes DK (1988) Pollination control: mechanical and sterility. pp. 931–960. *In*: Hanson AA, Barnes DK and Hill Jr RR (eds.). Alfalfa and Alfalfa Improvement. Madison, WI: American Society of Agronomy, Inc. Publishers.

Visarada KBRS, Padmaja PG, Saikishore N, Pashupatinath E, Royer M, Seetharama N and Patil JV (2014) Production and evaluation of transgenic sorghum for resistance to stem borer. *In Vitro* Cellular and Developmental Biology—Plant 50: 176–189.

vom Brocke K, Christinck A, Weltzien RE, Presterl T and Geiger HH (2003) Farmers' seed systems and management practices determine pearl millet genetic diversity patterns in semiarid regions of India. Crop Sci. 43: 1680–1689.

von Bothmer R and Jacobsen N (1985) Origin, taxonomy, and related species. pp. 19–56. *In*: Rasmusson DC (ed.). Barley. Madison, WI: American Society of Agronomy-Crop Science Society of America-Soil Science Society of America.

von Bothmer R and Komatsuda T (2011) Barley origin and related species. pp. 14–61. *In*: Ullrich SE (ed.). Barley Production, Improvement, and Uses. Oxford: Wiley-Blackwell.

Wagner DB and Allard RW (1991) Pollen migration in predominantly self-fertilizing plants: barley. J. Hered. 82: 302–304.

Waines JG and Ehdaie B (2007) Domestication and crop physiology: Roots of green-revolution wheat. Annals of Botany 100: 991–998.

Waines JG and Hegde SG (2003) Intraspecific gene flow in bread wheat as affected by reproductive biology and pollination ecology of wheat flowers. Crop Sci. 43: 451–463.

Warburton ML, Crossa J, Franco J, Kazi M, Rajaram S, Pfeiffer W, Zhang P, Dreisigacker S and van Ginkel M (2006) Bringing wild relatives back into the family: recovering genetic diversity in CIMMYT improved wheat germplasm. Euphytica 149: 289–301.

Warwick SI, Simard MJ, Legere A, Beckie HJ, Braun L, Zhu B, Mason P, Seguin-Swartz G and Stewart CN (2003). Hybridization between transgenic *Brassica napus* L. and its wild relatives: *Brassica rapa* L., *Raphanus raphanistrum* L., *Sinapis arvensis* L., and *Erucastrum gallicum* (Willd.) O.E. Schulz. Theor. Appl. Genet. 107: 528–539.

Warwick SI, Beckie HJ and Hall LM (2009) Gene flow, invasiveness, and ecological impact of genetically modified crops. Annals of New York Academy of Science 1168: 72–99.

Watabe T (1977) Rice Road (in Japanese). Tokyo: Nippon Housou Kyoukai.

Wegier A, Pineyro-Nelson A, Alarcon J, Galvez-Mariscal A, Alvarez-Buylla ER and Pinero D (2011) Recent long-distance transgene flow into wild populations conforms to historical patterns of gene flow in cotton (*Gossypium hirsutum*) at its centre of origin. Mol. Ecol. 20: 4182–4194.

Weiss EA (1983) Oilseed Crops. New York: Longman Group Ltd.

Wendorf F, Close AE, Schild R, Wasylikowa K, Housley RA, Harlan JR and Królik H (1992) Saharan exploitation of plants 8,000 years BP. Nature 359: 721–724.

Westengen OT, Huaman Z and Heun M (2005) Genetic diversity and geographic pattern in early South American cotton domestication. Theor. Appl. Genet. 110: 392–402.

Westphal E (1974) Pulses in Ethiopia: Their Taxonomy and Agricultural Significance. Agricultural Research Report 815. Wageningen, The Netherlands: Center for Agricultural Publishing and Documentation.

Whitton J, Wolf DE, Arias DM, Snow AA and Rieseberg LH (1997) The persistence of cultivar alleles in wild populations of sunflowers five generations after hybridization. Theor. Appl. Genet. 95: 33–40.

Wilkes HG (1977) Hybridization of maize and teosinte in Mexico and Guatemala and the improvement of maize. Econ. Bot. 31: 254–293.

Wittkop B, Snowdon RJ and Friedt W (2009) Status and perspectives of breeding for enhanced yield and quality of oilseed crops for Europe. Euphytica 170: 131–140.

Woodworth CM (1922) The extent of natural cross-pollination in soybeans. Agron. J. 14: 278–283.

Yamashita K (1980) Origin and dispersion of wheats with special reference to peripheral diversity. Z Pflanzenzucht. 84: 122–132.

Yan S, Zhu J, Zhu W, Li Z, Shelton AM, Luo J, Cui J, Zhang Q and Liu X (2015) Pollen-mediated gene flow from transgenic cotton under greenhouse conditions is dependent on different pollinators. Scientific Reports 5: 15917, DOI: 10.1038/srep15917.

Ye X, Al-Babili S, Klöti A, Zhang J, Lucca P, Beyer P and Potrykus I (2000) Engineering the provitamin A (β-carotene) biosynthetic pathway into (carotenoid-free) rice endosperm. Science 287: 303–305.

Yoshimura Y, Matsuo K and Yasuda K (2006) Gene flow from GM glyphosate-tolerant to conventional soybeans under field conditions in Japan. Environ. Biosafety Re. 5: 169–173.

Zhao R, Bean S, Wu X and Wang D (2008) Assessing fermentation quality of grain sorghum for fuel ethanol production using rapid visco-analyzer. Cereal. Chem. 85: 830–836.

Zohary D and Hopf M (1988) Domestication of plants in the old world: The Origin and Spread of Cultivated Plants in West Asia, Europe, and the Nile Valley. Oxford: Clarendon Press. 328 p.

Zohary D, Hopf M and Weiss E (2012) Domestication of Plants in the Old World: The Origin and Spread of Domesticated Plants in Southwest Asia, Europe, and the Mediterranean Basin. Oxford: Oxford Univ. Press. pp. 51–59.

Chapter 7

Crop Selectivity and Herbicide Safeners

Historical Perspectives and Development, Safener-Regulated Gene Expression, Signaling, and New Research Directions

Dean E. Riechers[1] and Jerry M. Green[2,]*

Introduction to Herbicide Safeners

Herbicide safeners are chemical compounds used to protect monocot crops (typically large-seeded cereals) from herbicide injury. Safeners are chemically diverse and possess the unique ability to selectively protect cereal crops from herbicide injury, which is accomplished by inducing metabolic detoxification reactions (Kraehmer et al. 2014; Riechers et al. 2010; Rosinger et al. 2012). Herbicide safeners are flexible in their application method (e.g., crop seed treatment, preemergence (PRE) with the herbicide, or postemergence (POST) with the herbicide) and are novel crop protection agents in that they only protect cereal crops from herbicide injury (Hatzios 1983, 1991; Parker 1983). Perplexingly, commercial safeners do not confer significant phenotypic effects (i.e., prevention of herbicide injury) in dicot crops or in most weed species (Jablonkai 2013). Most herbicide safeners used today act by stimulating herbicide detoxification mechanisms in cereal crops (Breaux et al. 1989; Cole 1994; Fuerst et al. 1986; Hatzios and Burgos

[1] Professor, Department of Crop Sciences, University of Illinois, Urbana, IL 61801.
[2] President, Green Ways Consulting LLC, Landenberg, PA 19350.
* Corresponding author: riechers@illinois.edu

2004; Riechers et al. 2010), although some examples of "inactive" herbicide antagonists may also be classified as safeners, which will be discussed further below. Chemicals have not been commercialized solely to safen broadleaf crops from herbicides, in spite of ample evidence that safeners induce gene expression in *Arabidopsis thaliana* (DeRidder et al. 2002, 2006; De Veylder et al. 1997; Hershey and Stoner 1991).

Herbicide safeners tap into pre-existing signaling pathways to trigger expression of genes involved in plant defense and detoxification, such as glutathione *S*-transferases (GSTs) and cytochrome P450s (Cummins et al. 2011; Riechers et al. 2010; Skipsey et al. 2011). In addition to their ability to selectively protect cereal crops from herbicide injury, safeners are also unique and valuable tools for studying early signaling and stress-response genes, the tissue-specific regulation of GST and P450 expression, and for inducing the expression of essential components of a coordinated detoxification pathway (Theodoulou et al. 2003; Zhang et al. 2007) in cereal crops in the absence of phytotoxicity. The objectives of this chapter are to review the historical development and commercialization of herbicide safeners and provide an update on recent advances in our understanding of safener-mediated signaling pathways for regulating plant defense gene expression in a tissue-specific manner. Several recent articles have reviewed the biochemical effects of safeners on detoxification enzyme activities and herbicide metabolism pathways in cereal crops (Hatzios and Burgos 2004; Jablonkai 2013; Rosinger et al. 2012); as a result, these topics will not be addressed in detail in this chapter.

Discovery of the First Herbicide Safener

Herbicide safeners were discovered inadvertently by an astute scientist under fortuitous circumstances. In the summer of 1947, shortly after 2,4-D was discovered, Otto Hoffmann was involved in a test that treated a greenhouse full of tomato (*Lycopersicon esculentum* L.) plants with 2,4-D analogs. The weather was warm and the greenhouse had not been adequately ventilated during the weekend. 2,4-D and some of its analogs were volatile and caused severe damage to tomato plants throughout the greenhouse, almost all the tomato plants. Before the test was thrown out, Otto Hoffman viewed the disaster and noticed the one plant treated with 2,4,6-T, was not injured. He perceived its significance and his follow-up studies showed that 2,4,6-T was not herbicidal but that it could prevent the herbicidal effect of 2,4-D by competitive binding at its active site (Hansch and Muir 1950; Hoffmann 1953). He called this new chemical class "herbicide antidotes." Due to that discovery and for devoting his career to develop the use of non-herbicidal chemicals to enhance crop tolerance, Otto Hoffmann has been called the "father of herbicide antidotes." Today, the term herbicide antidote has fallen out of favor, primarily because of the use of antidote in medicine to reverse unwanted chemical effects. Herbicide antidotes do not generally reverse herbicidal effects, but rather prevent them. Herbicide safener is now the preferred terminology (Davies and Caseley 1999; Hatzios 1983; Kraehmer et al. 2014; Parker 1983).

Definition and Need for Safeners

Herbicide safeners greatly assist with achieving crop safety and the required margin of crop-weed selectivity, which historically has been the most difficult aspect of herbicide discovery to achieve until the advent of genetically modified herbicide-resistant crops (Green 2014). Figure 1 displays how safeners shift the crop-response curve of a given herbicide. The greater the shift to the right equates to a greater reduction in herbicide activity, which results in enhanced crop tolerance. Herbicide safeners can only be successful commercially when they consistently and significantly reduce crop injury but do not reduce the herbicide activity on key target weeds (Green and Amuti 1993; Jablonkai 2013).

Herbicide safeners are more important now than ever before (Jablonkai 2013; Kraehmer et al. 2014; Rosinger 2015) despite the widespread adoption of transgenic, herbicide-resistant (HR) crops where herbicide safeners are not needed to provide crop tolerance to non-selective herbicides with limited crop metabolism (Green 2014). Herbicide safeners have several distinct advantages in acting as chemical inducers for enhanced crop

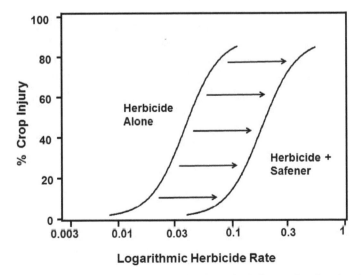

Figure 1. Logarithmic dose-response curve with a hypothetical crop showing the effect of a herbicide safener, which reduced the activity of the herbicide about 8-fold.

tolerance. Safeners may permit the selective control of weeds in botanically related crops without the use of special, typically expensive, transgenic varieties (Jablonkai 2013). Safeners may also expand the utility of existing herbicides that do not possess adequate crop safety when applied without a safener, which are desperately needed to manage multiple-resistant weeds (Yu and Powles 2014). Furthermore, safeners increase the chance of successful herbicide discovery by giving researchers a tool to overcome a common discovery hurdle: achieving sufficient crop tolerance and/or crop-weed selectivity (Duke 2012). In addition to their important roles in crop protection, herbicide safeners have other interesting utilities. For example, the use of the safener *N*-(aminocarbonyl)-2-chlorobenzenesulfonamide, or 2-CBSU (De Veylder et al. 1997; Hershey and Stoner 1991) has been patented to activate gene expression at specific times of plant development stages (Cigan and Unger-Wallace 2015); thus, the safener acts as an external chemical "gene switch" (Padidam 2003; Ward et al. 1993) that can be utilized to precisely manipulate transgene expression in crops.

Definitions can become confusing when agrochemicals that are used for other purposes are used as herbicide safeners. When herbicides are used at low rates (without herbicidal activity) to reduce crop injury, they are technically being utilized as safeners. The same can be said about insecticides and other types of agrochemicals that are considered herbicidally inactive by themselves (Green 1989). However, if the herbicide is used at a labeled rate and still retains herbicidal activity on target weeds but also safens the crop from injury, it is considered herbicide-herbicide antagonism and therefore is not classified as a herbicide safener (Green 1989; Green and Amuti 1993).

Examples of herbicides that have been used widely as herbicide safeners include diamuron, a urea herbicide, which safens rice from bensulfuron, as well as cumyluron and dimepiperate (Rosinger et al. 2012). Examples of other pesticides currently used to safen from herbicides are the organophosphate (OP) insecticides, particularly disulfoton and phorate, which are used to safen cotton from clomazone. The clomazone label (Anonymous 2013) requires these insecticides to be applied in-furrow when planting cotton. OP insecticides inhibit P450-mediated metabolism and prevent the inactive, proherbicide clomazone from metabolic conversion (i.e., bioactivation) to its active form, 5-keto-clomazone (Ferhatoglu et al. 2005). Similarly, the non-phytotoxic microbial OP inhibitor dietholate [*O,O*-diethyl-*O*-phenyl phosphorothioate] (Tam et al. 1988) inhibits soil microbes that degrade thiocarbamate herbicides, and was recently patented to safen cotton from clomazone injury (Keifer 2005).

Finally, activated carbon or charcoal can also be considered a herbicide safener (Hatzios and Hoagland 1989). Activated carbon is a non-herbicidal, porous, soft, black substance made by heating carbon materials that can absorb 100 to 200 times its weight of organic chemicals, including herbicides. This general inactivation method is useful to cleanup persistent environmental chemical spills and has been studied for use in cropping situations (Yelverton et al. 1993), but with little commercial use to date.

Discovery, Development, and Commercialization of Herbicide Safeners

Otto Hoffman discovered the first commercial safener, naphthalic anhydride (NA) (Hatzios and Hoagland 1989). NA was commercialized in 1969 under the tradename Protect® and was fairly successful for a decade. After NA had been discontinued, Otto Hoffmann described NA as being partially able to safening from most herbicides most ways it was applied, but never achieved total crop safening (Jablonkai 2013). Incomplete safening from crop injury, particularly severe injury, is still an issue for developing new safeners (Jablonkai 2013; Kraehmer et al. 2014).

The discovery of herbicide safeners has generally been accomplished through random chemical screening, or by directed synthesis and testing chemical analogs derived from the herbicide of interest. Hence, many herbicide safeners appear similar as herbicides (Komives and Hatzios 1991; Rosinger 2012). For example, Hatzios (1983, 1991) and Hatzios and Hoagland (1989) have described several examples of how the optimization process has progressed through several generations of herbicide safener products.

An early observation offering crucial insight into one of the first classes of herbicide safeners was that low doses of allidochlor (*N,N*-diallyl-2-chloroacetamide), an acetamide herbicide also known as CDAA and Randox®, applied 1 to 2 days before a 40-fold higher rate of allidochlor significantly reduced corn injury (Ezra et al. 1985). This herbicide pretreatment increased GST enzyme activity by 61% within one day. A similar phenomenon occurred when pretreating corn with soil-applied atrazine, which safened the corn from subsequent PRE atrazine applications (Jachetta and Radosevich 1981). Interestingly, the safener dichlormid is only one chlorine atom different than allidochlor and also performs this metabolic function of safening corn (Lay et al. 1975; Lay and Casida 1976, 1978). Dichlormid was commercialized in 1971 as the first member of the dichloracetamide family of safeners (Table 1), which has at least eight commercial members (Hatzios and Hoagland 1989).

The list of commercial safeners is fairly long, particularly for corn (*Zea mays* L.), wheat (*Triticum aestivum* L.), rice (*Oryza sativa* L.) and grain sorghum (*Sorghum bicolor* (Moench) L.) (Table 1). The chemical structures for two representative safeners from this list, fluxofenim and cloquintocet-mexyl, are depicted in Fig. 2. Initially, herbicide safeners were not considered

Table 1. Chronological list of important commercial herbicide safeners and the key crop they safen. The last four safeners included (listed without commercialization dates) were developed in Japan for rice.

Herbicide Safener	First Sales	Company	Key Crop
Naphthalic anhydride (NA)	1969	Gulf Oil	Multiple Crops
Dichlormid	1971	Stauffer	Corn
R-29148	1973	Stauffer	Corn
Cyometrinil	1978	Ciba-Geigy	Sorghum
Flurazole	1980	Monsanto	Sorghum
Oxabetrinil	1982	Ciba-Geigy	Sorghum
Fenclorim	1983	Ciba-Geigy	Rice
BAS-145138	1984	BASF	Corn
Fluxofenim	1986	Ciba-Geigy	Sorghum
MG-191	1986	Nitrokemia	Corn
Benoxacor	1988	Ciba-Geigy	Corn
Cloquintocet-mexyl	1989	Ciba-Geigy	Wheat
Fenchlorazole-ethyl	1989	Hoechst	Wheat
Furilazole	1992	Monsanto	Corn, Wheat
Mefenpyr-diethyl	1997	AgrEvo	Wheat, Barley
Isoxadifen-ethyl	2000	Aventis	Corn, Rice, Wheat
Cyprosulfamide	2009	Bayer	Corn
Dimepiperate			Rice
Diamuron			Rice
S-PEU			Rice
Cumyluron			Rice

active ingredients and were not regulated nearly as much as herbicides. Hence, the cost to develop herbicide safeners was much less, and as a result smaller market opportunities could be targeted. Currently, herbicide safeners are not required to be listed as active ingredients, but companies usually disclose them included on label, the corresponding safety data sheet, and/or in their promotional literature.

As mentioned previously, herbicide safeners often appear structurally similar as herbicides (Hatzios and Hoagland 1989; Jablonkai 2013; Lay et al. 1975) and are now increasingly regulated. In fact, some agrochemical companies currently consider the information needed to commercialize a safener or herbicide to be very similar (Kraehmer et al. 2014; Rosinger et al. 2012), so the associated development costs are high and safeners must now address large commercial opportunities in order to be economically viable.

Herbicide safeners must function in what the popular press is calling the age of herbicide mixtures. Today, crop-selective herbicides are rarely

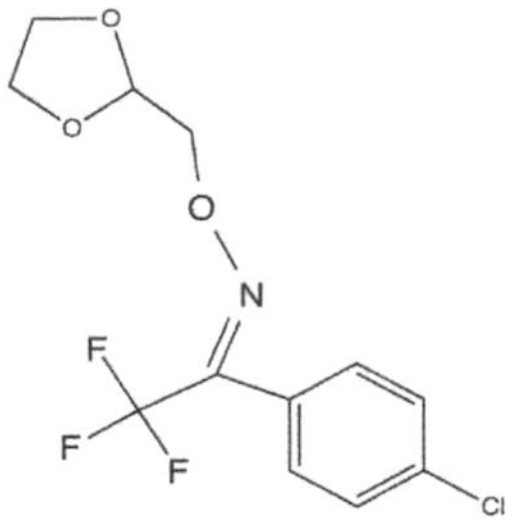

(*Z*)-1-(4-chlorophenyl)-2,2,2-trifluoroethan-1-one *O*-((1,3-dioxolan-2-yl)methyl) oxime

Fluxofenim (Concep III)

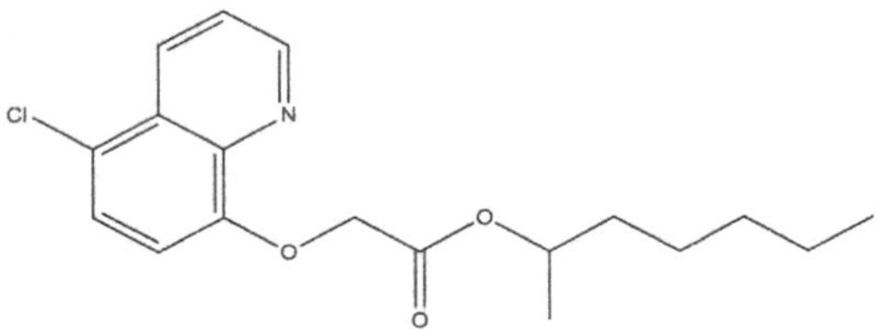

1-Methylhexyl (5-chloroquinolin-8-yloxy)acetate

Cloquintocet-mexyl

Figure 2. Representative herbicide safeners used in grain sorghum (fluxofenim; seed treatment) and wheat (cloquintocet-mexyl; postemergence).

used alone (Green 1991). The large array of herbicides being applied with herbicide safeners is putting tremendous pressure on safeners to only safen crops and not the targeted weeds. Such crop-specific safening is changing the herbicide safener discovery process from the need to safen from specific herbicides to safening specific crops. As a consequence, new safeners tend to be designed to focus on unique metabolic interactions of safeners and herbicides in specific crops (Rosinger 2015). Co-discovery of herbicides and safeners may assist in this challenging endeavor, as well as discovering and developing prosafeners, which are inactive in their parent form but become effective herbicide safeners upon metabolic bioactivation in the crop (Jeschke 2016). The concept of developing prosafeners for herbicides will be discussed in more detail in the next section.

Currently, many of the new herbicide products for use in cereal crops have safeners (Table 1), particularly in rice, grain sorghum, and winter wheat where transgenic HR traits are not available due to the presence of wild, weedy relatives. An interesting trend started in the early 1990s, when more POST safeners were discovered and commercialized with POST herbicides (Table 1), and continues today (Kraehmer et al. 2014). This trend coincided with the use of high unit-activity herbicide molecules (Jablonkai 2013), increased knowledge of the roles of spray adjuvants in enhancing herbicide uptake on plant foliage, and rapid grower adoption of total POST weed management programs. More recent utilities of existing safener and herbicide chemistry are to: (1) broaden the application-timing window for a herbicide previously limited to PRE applications, as is the case for the safener cyprosulfamide (Table 1) in combination with isoxaflutole (Philbrook and Santel 2008), or (2) protect corn from herbicides with differing sites-of-action with a single safener, such as isoxadifen-ethyl (Table 1) (Jablonkai 2013; Schulte and Köcher 2009). The combination of cyprosulfamide plus isoxaflutole allows growers to apply the herbicide after corn emergence (Rosinger et al. 2012; Kraehmer et al. 2014), which previously could not be accomplished due to high levels of crop injury, thereby providing flexibility in application timing and extending the effective period for residual weed control.

Propesticides and Discovering New Safeners

Numerous pesticides commercialized today, including insecticides and herbicides, are not actually toxic in the form that is applied to the pest. Instead, these compounds called "propesticides" must be converted into the active form by enzymes within the insect or plant (Jeschke 2016). This process, also called bio-activation, has been used successfully to commercialize crop-selective herbicides, such as the 4-hydroxyphenyl-pyruvate dioxygenase (HPPD)-inhibiting diketonitrile-derivative of the herbicide isoxaflutole (Pallett et al. 2001). Benefits of utilizing proherbicides is that the inactive form is often more suitable from a biokinetic perspective, such as more efficient or rapid foliar absorption by the target weed species, as well as potential for more precise modulation of the biological half-life of the herbicide *in planta* or in soil (Jeschke 2016). In regards to the concept of prosafeners (Jablonkai 2013), one strategy used successfully in several commercial safeners is esterification of carboxylic acids (Jeschke 2016). As noted in Table 1, several of the common POST safeners are ethyl (or diethyl for mefenpyr) esters of the corresponding parent carboxylic acid. A unique methylhexyl or "mexyl" group was used to esterify the POST safener cloquintocet acid (Fig. 2). Esterification chemistry has been used to synthesize both proherbicides and prosafeners, and rely on hydrolytic esterases located in the leaf cuticle or cell wall to rapidly bioactivate the ester to the free acid (Taylor et al. 2013). In addition to improved foliar absorption of the prosafener, upon hydrolysis of the ester group the safener acid becomes systemic (Jeschke 2016) and moves in a source-to-sink manner throughout the plant in the phloem. In the case of foliar-applied cloquintocet-mexyl and cloquintocet acid, the acidic form of cloquintocet was equally effective as the esterified form at inducing GST activities in wheat foliage through 48 hr after treatment, suggesting that cloquintocet acid is the actual safening molecule *in planta* (Taylor et al. 2013).

In theory it may be possible to chemically engineer synchronous or asynchronous (i.e., temporally separated) uptake and translocation of a POST herbicide in a tank-mix combination with a POST safener, as well as optimizing adjuvant selection. For example, it may be beneficial for enhancing crop tolerance to have an asynchronous uptake pattern that allows the prosafener to enter plant cells first to trigger defense and detoxification enzymes for "priming" rapid metabolism before the proherbicide (or active herbicidal acid) enters the cells.

Tissue- and Organ-Specific Expression of GSTs following Safener Treatment

A limited number of studies have investigated the distribution of safener-induced GST proteins in plant organs and tissues (Holt et al. 1995; Riechers et al. 2003; Sari-Gorla et al. 1993; Taylor et al. 2013). Unfortunately, however, none of these studies have examined the tissue distribution of other essential enzymes that comprise the entire herbicide detoxification pathway in plants. Prior to these GST tissue-distribution studies, either crude protein, total RNA, or cell-free plant extracts had been utilized for determination of GST activities, protein abundance, or transcript levels within whole seedlings, roots, or shoots with little consideration of tissue-specific localization before or after safener treatment (Riechers et al. 2005). However, recent findings have shown that several different GST isoforms (both constitutive and safener-inducible) are expressed in the various plant tissues and organs examined in these studies, suggesting a complex developmental regulation of GST expression in plants. For example, a recent study investigated tissue-specific expression of various GST protein subclasses in the foliage of untreated and safener-treated wheat seedlings (Taylor et al. 2013) by dividing wheat seedlings into leaves and meristematic tissues, and the leaves further sub-divided into leaf tips and middle sections. GST activity in untreated control plants was four-fold higher in meristems than in the foliage, and the leaves were more responsive to safener treatment than meristems (Taylor et al. 2013). Moreover, as determined by immunoblotting, both phi and tau class GSTs were induced by cloquintocet-mexyl in all tissues examined (most noticeably in the leaves for the phi-class GSTs), whereas safener induction of lambda-class GST enzymes was mainly limited to the meristems (Taylor et al. 2013). Future research examining GST tissue-specific expression patterns of individual GST isoforms in the foliage of additional cereal crops will help to clarify their precise roles in herbicide metabolism and safener-induced detoxification pathways.

In contrast with cloquintocet-mexyl and other recently developed foliar-applied safener-herbicide combinations, many of the first safeners introduced into agriculture were developed for herbicides of the chloroacetamide herbicide classes for maize, grain sorghum, and rice, which are typically applied together with a safener to the soil (at or before planting) or as seed treatments (Table 1). These herbicides are primarily detoxified in tolerant crop plants by GST-catalyzed conjugation with reduced glutathione (GSH) (Hatzios and Burgos 2004). The main location of herbicide uptake is the shoot coleoptile, while the most sensitive sites are the shoot meristems and developing leaves within (Riechers et al. 2010). When applying the chemically unrelated safeners cloquintocet-mexyl and fluxofenim to shoots of the diploid wheat species *Triticum tauschii*, large increases in GST transcript, protein, and specific enzyme activity levels were detected mainly in the coleoptile tissues (Riechers et al. 2003). Additionally, a dramatic increase in the level of immuno-reactive GST protein was observed specifically in the outermost cell layers of the coleoptile (Riechers et al. 2003). It follows that safeners for chloroacetamides appear to exert their main effect on herbicide metabolism in the coleoptiles of etiolated shoots by inducing GSTs that rapidly detoxify the herbicide as the developing shoot and leaves emerge from the soil (Kreuz et al. 1989; Riechers et al. 2003). These findings are in agreement with chloroacetamide metabolism studies in maize seedlings, where benoxacor (Table 1) increased GSH-mediated metabolism of metolachlor mainly in maize coleoptiles (Kreuz et al. 1989).

Collectively, these studies have established that expression of many different GST subunits and isoforms may be organ- or tissue-specific, developmentally regulated, and that GST genes from various subclasses (Frova 2003, 2006; Mashiyama et al. 2014) differentially respond to safeners (Holt et al. 1995; Taylor et al. 2013). Regarding the regulation of entire cellular detoxification pathways by safeners (Zhang et al. 2007), the coordinate induction of Phase II detoxification reactions (via GSH or glucose conjugation) and Phase III vacuolar transport mechanisms (Theodoulou et al. 2003), and co-localization of these mechanisms to the outer layers of the coleoptile would allow for herbicide detoxification to occur at the site near shoot uptake (Riechers et al. 2005, 2010). Other constitutive and inducible metabolic enzymes co-induced by safener treatment (such as P450s) may also be localized in relation to which GST isoforms are present in various plant parts, but information on the subcellular localization and tissue distribution of these safener-induced proteins is scarce in the literature.

Mechanisms of Safener-Regulated Gene Expression in Plants

Plants are frequently exposed to synthetic toxins, abiotic stresses, or endogenous, reactive electrophilic species (RES) that elicit the rapid production of reactive oxygen species (ROS) and activation of plant defense mechanisms for adaptation and survival (Almeras et al. 2003; Baxter et al. 2014; Farmer and Davoine 2007). As mentioned previously, herbicide safeners are non-phytotoxic compounds that confer protection from herbicide injury in cereal crops by inducing detoxification systems for endogenous toxins, xenobiotics, and ROS (Riechers et al. 2010). Safeners tap into pre-existing signaling pathways to promote the expression of detoxification genes such as GSTs and P450s, yet these signaling pathways remain poorly defined. However, safeners provide unique and valuable tools for studying early signaling and stress-response genes, the regulation of GST and P450 gene expression, and for inducing the expression of other essential components of the well-documented three-phase detoxification pathway (Kreuz et al. 1996; Theoudoulou et al. 2003; Riechers et al. 2010) in cereal crops.

Recent studies have indicated that safeners induce the expression of GSTs that detoxify PRE herbicides mainly in the outer three cell layers (i.e., epidermal and sub-epidermal) of cereal crop seedling coleoptiles (Riechers et al. 2003, 2010). These findings have led to new hypotheses that (a) safeners are tapping into an unidentified, pre-existing signaling pathway for plant defense and detoxification in a *tissue-specific* manner, and (b) safeners may be utilizing an oxidized lipid ('oxylipin' or cyclopentenone; Fig. 3)—mediated signaling pathway (Mueller 2004; Dave and Graham 2012; Riechers et al. 2010) in the coleoptile, which subsequently leads to dramatic but specific induction of plant defense genes involved with detoxification in epidermal and sub-epidermal cell layers. Several theories for describing the signaling mechanisms triggered by safeners have been postulated, and will be discussed further below.

Potential models for abiotic stress signaling in response to xenobiotics have been proposed, but none have been clearly tested using safeners (Irzyk and Fuert 1997; Hatzios and Burgos 2004; Ramel et al. 2012). One confounding factor that hinders safener-signaling research is that it remains difficult to distinguish between contrasting hypotheses of direct xenobiotic sensing and indirect sensing of xenobiotic-related modifications or cellular events (Ramel et al. 2012). In one proposed model, xenobiotic-induced gene expression may result from oxidative stress generated within the cell. Possible elicitors include an alteration in the reduced glutathione to oxidized glutathione ratio (GSH:GSSG), indicating perturbed glutathione homeostasis (Farago et al. 1994; Ramel et al. 2012), or the production of ROS (Baxter et al. 2014). Safener-GSH conjugates have been identified in *Arabidopsis* and maize (Brazier-Hicks et al. 2008; Liu et al. 2009; reviewed by Riechers et al. 2010), although it is not known if this is non-enzymatic or GST-catalyzed GSH conjugation of the parent safener molecule. Detailed information regarding the biochemical and molecular events that occur between the initial safener application and the end result (i.e., increased GST activity, enhanced herbicide metabolism, and increased crop tolerance) in safener-responsive cereal crops is limited. However, possible mechanisms for safener-regulated signaling mechanisms involving oxylipins (Fig. 3)

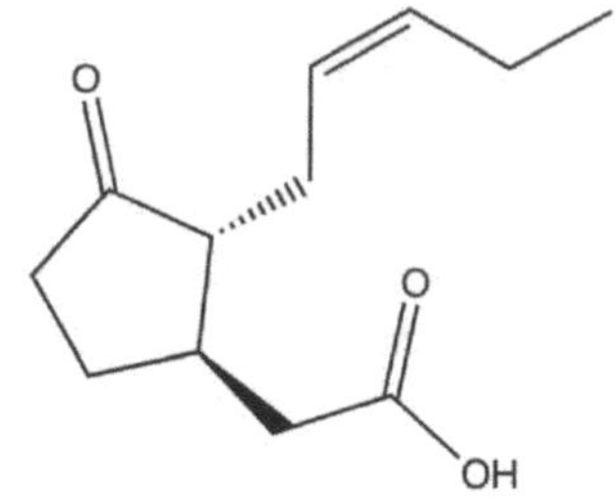

(1R,2R)-3-oxo-2-(2Z)-2-pentenyl-cyclopentaneacetic acid

Jasmonic acid (**JA**)

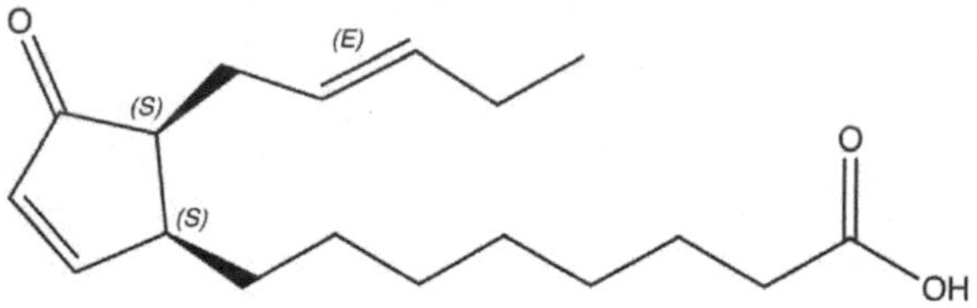

8-((1*S*,5*S*)-4-oxo-5-((*E*)-pent-2-en-1-yl)cyclopent-2-en-1-yl)octanoic acid

12-oxo-phytodienoic acid (**OPDA**)

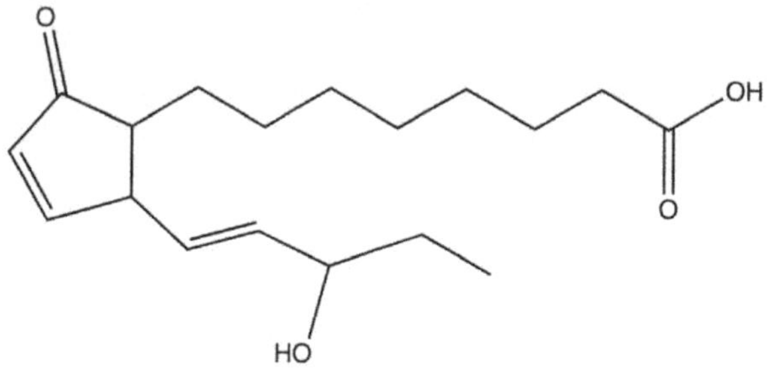

2-(3-hydroxy-1-penten-1-yl)-5-oxo-3-cyclopentene-1-octanoic acid

Phytoprostane-A$_1$ (**PPA$_1$**)

Figure 3. Chemical structures of oxidized fatty acids/lipids (oxylipins) and jasmonic acid (JA), which function as signaling molecules in plant defense and detoxification reactions. Note that JA contains a cyclopentanone ring while OPDA and PPA$_1$ contain cyclopentenone rings.

have been proposed (Riechers et al. 2010; Skipsey et al. 2011), as has the involvement of TGA and WRKY transcription factors (Behringer et al. 2011; Stotz et al. 2013). The potential roles of oxylipins in safener-regulated signaling mechanisms will be described in more detail below.

Structures, Synthesis, and Roles of Oxylipins in Plant Defense Signaling

Oxylipins are structurally diverse metabolites derived from fatty acid oxidation, and can be formed through either non-enzymatic or enzymatic reactions (Fig. 3). Non-enzymatically generated oxylipins are formed via free radical-catalyzed reactions in or near cell membranes, where polyunsaturated fatty acids (such as α-linolenic acid; Christeller and Galis 2014) serve as precursors for their synthesis, and include many different types of phytoprostanes, malondialdehyde, and 4-hydroxy-2E-nonenal (Mueller 2004; Mosblech et al. 2009). Enzymatically produced oxylipins include jasmonic acid (JA) and 12-oxo-phytodienoic acid (OPDA); this pathway has been well studied due to the hormonal activity of JA and defense gene activation (Stintzi et al. 2001; Mosblech et al. 2009; Schaller and Stintzi 2009; Yan et al. 2012; Christensen et al. 2014). JA is synthesized via a series of steps starting with the oxygenation of α-linolenic acid from membrane lipids via lipoxygenase activity (Vicente et al. 2012; Grebner et al. 2013; Zhou et al. 2014) and subsequent conversion to OPDA (Schaller and Stintzi 2009; Yan et al. 2012).

Oxylipins differ not only in their subcellular origin, synthesis, and structures (Fig. 3) but also in their degree of electrophilicity. For example, strong RES include the A_1-type phytoprostanes and OPDA, whereas relatively weak RES include the B_1-type phytoprostanes and JA (Farmer and Davoine 2007). Since cyclopentenones (i.e., phytoprostanes and precursors of JA) and cyclopentanones (i.e., JA) differ in their biological activities (Stintzi et al. 2001; Taki et al. 2005), the reduction of the cyclopentenone ring by 12-oxo-phytodienoic acid reductases (OPRs) (Fig. 3) may therefore be critical for regulation of the entire octadecanoid pathway, as this enzymatic step controls the relative levels of these two classes of signaling molecules (Schaller and Stintzi 2009; Christensen et al. 2014).

Lipase-induced release of specific lipid substrates (such as α-linolenic acid) in response to abiotic and biotic stresses enables their rapid conversion into various classes of oxylipins, which may perceive and respond to a wide range of environmental stimuli (Almeras et al. 2003; Bonaventure et al. 2011; Christeller and Galis 2014; Schuck et al. 2014). Membranes in the epidermal and sub-epidermal cell layers of plant tissues and organs are likely the initial sites of perception for a wide range of environmental stressors (Javelle et al. 2011; Okazaki and Saito 2014), and possibly herbicide safeners as well (Riechers et al. 2003, 2005). Polyunsaturated fatty acids are major structural constituents of cell membranes that also function as modulators of diverse signal transduction pathways triggered by abiotic stresses (Okazaki and Saito 2014; Savchenko et al. 2014). Different stresses induce the production of different classes of oxylipins that regulate distinct, yet partially overlapping, transcriptional responses (Taki et al. 2005; Mueller et al. 2008; Schuck et al. 2014). Recent results suggest that enzymatically-formed oxylipins (e.g., OPDA and jasmonates) and non-enzymatically formed oxylipins (e.g., phytoprostanes) perform important but distinct functions in plant defense responses (Mueller and Berger 2009; Dave and Graham 2012; Savchenko et al. 2014).

Common genes and proteins induced by both safeners and various classes of oxylipins include several P450s, uGTs, GSTs, and glutathione-conjugate ABC transporters (Loeffler et al. 2005; Taki et al. 2005; Dueckershoff et al. 2008; Mueller et al. 2008; Riechers et al. 2010). Interestingly, the majority of these genes and proteins are also induced in *Arabidopsis* by allelochemical treatment (Baerson et al. 2005). However, the specific roles of different classes of oxylipins in plant defense and detoxification mechanisms, most notably their roles in xenobiotic sensing and signaling in cereal crops, remain unclear. The possibility that OPDA, JA, JA-isoleucine, phytoprostanes, and other oxidized fatty acids act in concert *in planta* to precisely regulate defense gene expression (Mueller and Berger 2009) in different plant cells, tissues, or organs has not been investigated in detail but warrants further inspection.

Knowledge Gaps in Mechanisms of Safener-Mediated Signaling in Cereal Crops

In spite of the wealth of physiological, biochemical, and phenotypic information regarding the use of safeners to protect cereal crops from herbicides (Davies and Caseley 1999; Hatzios and Burgos 2004; Jablonkai 2013), there is comparatively little known regarding safener-mediated signaling pathways in safener-responsive cereal crops (Riechers et al. 2010). Most of the information published on safener-regulated signaling mechanisms is derived from studies conducted in phenotypically safener-unresponsive model dicots. These studies have included the model plant *Arabidopsis* (Behringer et al. 2011) and *Populus* (Rishi et al. 2004), particularly with regard to describing the roles of TGA/WRKY transcription factors (Behringer et al. 2011) and oxylipins (Skipsey et al. 2011). Root cultures from wild type *Arabidopsis* plants and mutants defective in fatty acid desaturation (*fad3-2/fad7-2/fad8*), which are defective in forming the oxylipin precursor α-linolenic acid, demonstrated a decreased ability to respond to safener treatment when measuring *AtGSTU24* expression (Skipsey et al. 2011). Since these *fad* mutants accumulate linoleic acid instead of α-linolenic acid, they are unable to synthesize OPDA or phytoprostanes from α-linolenic acid substrates released via lipase activities (Christeller and Galis 2014). The decreased ability of these mutant *Arabidopsis* lines to respond to safener treatment (measured by the lack of induction of GST expression) is consistent with a link between safener-regulated defense responses and endogenous oxylipin signaling. As a result, the attenuated GST induction by safener treatment in *Arabidopsis* mutants displaying a reduction in polyunsaturated fatty acids suggests that safeners must act either in parallel or upstream of oxylipin signaling, potentially through regulating the availability of these endogenous molecules via hydrolytic lipase activities (Skipsey et al. 2011; Okazaki and Saito 2014). A unifying model can now be postulated that integrates the roles of lipase activities, α-linolenic acid, and the various classes of oxylipins in safener-regulated signaling of plant defense gene expression and detoxification reactions in cereal crops (Riechers et al. 2010).

Conclusion

Summary and future directions

In light of new findings related to safener mechanism of action in cereal crops (Riechers et al. 2003, 2010; Taylor et al. 2013; Zhang et al. 2007), a new theory resulting from previous research is that safeners may be utilizing an oxylipin-mediated signaling pathway, which subsequently leads to the expression of GSTs and other proteins involved in detoxification responses (Loeffler et al. 2005; Mueller et al. 2008; Mueller and Berger 2009). Oxylipins are constitutively present in higher plants, and as documented within the past decade, their levels increase in response to a wide variety of abiotic and biotic stresses (Christensen et al. 2014; Mosblech et al. 2009; Mueller 2004; Mueller and Berger 2009). As a result, a working hypothesis is that safeners trigger a massive release of free membrane lipids via lipase activity in the coleoptile, these lipids are then non-enzymatically converted to phytoprostanes by radical-mediated reactions, and finally the outermost coleoptile cells respond to the massive accumulation of phytoprostanes and attempt to restore homeostasis by expressing OPRs to reduce (Mueller et al. 2008), and GSTs to conjugate (Dixon and Edwards 2009; Dixon et al. 2010), the reactive, unstable phytoprostanes. However, a free pool of phytoprostanes might remain in the cell that subsequently triggers more expression of OPRs, GSTs, P450s, and other detox proteins to alleviate the oxidative stress generated in the outer coleoptile layers. It is plausible that at least some of these up-regulated GSTs and P450s can recognize herbicides (or other xenobiotics) as substrates, thus conferring a "coincidental" safening phenotype to the cereal crop seedling.

GST protein localizations in safener-treated wheat (Riechers et al. 2003) and sorghum shoot coleoptiles (Ma et al. unpublished results) indicate that highly-conserved, inducible signaling and detoxification pathways are present in cereal crop seedling tissues (Theodoulou et al. 2003; Zhang et al. 2007). Importantly, future studies investigating tissue- or cell-specific patterns of gene expression should also aim to determine the subcellular locations of detoxification proteins before and after treatment with safeners (Riechers et al. 2003, 2005). Additionally, temporal patterns of gene expression in response to safeners would be informative from the standpoint of understanding the time course of plant defense and detoxification responses following safener treatment in etiolated shoots or plant foliage. Elucidating the number, distribution, and potential coordinated regulation of safener-induced GST isoforms and other herbicide-detoxifying mechanisms across cells, tissues, and organs will likely provide new insights into GST function and regulation (Riechers et al. 2005; Taylor et al. 2013). An important question that needs to be experimentally addressed is whether GST protein localization is determined by its gene sequence (or encoded amino acid sequence), or if its subcellular localization is a direct result or consequence of the detoxification process it catalyzes (Riechers et al. 2005). Future research using proteomic techniques (Zhang and Riechers 2004), subcellular localization methods, RNAseq transcript analysis, and lipid/oxylipin profiling (Mueller et al. 2006; Christensen et al. 2014) will further investigate the hypothesis that safeners protect cereal crops from herbicide injury (or other abiotic stresses) by utilizing similar or parallel oxylipin-mediated signaling pathways for plant defense, including the cell- and tissue-specific coordinate induction of genes encoding herbicide detoxification enzymes.

Acknowledgement

The first author acknowledges funding provided by the Agriculture and Food Research Initiative, Competitive Grant #2015-67013-22818 of the USDA National Institute of Food and Agriculture, to support the research conducted in his laboratory program.

LITERATURE CITED

Almeras E, Stolz S, Vollenweider S, Reymond P, Mene-Saffrane L and Farmer EE (2003) Reactive electrophile species activate defense gene expression in *Arabidopsis*. Plant J. 34: 205–216.

Anonymous (2013) Command® 3ME herbicide product label. FMC Corporation. Philadelphia, PA: FMC. 19 p.

Baerson SR, Sanchez-Moreiras A, Pedrol-Bonjoch N, Schulz M, Kagan IA, Agarwal AK, Reigosa MJ and Duke SO (2005) Detoxification and transcriptome response in *Arabidopsis* seedlings exposed to the allelochemical benzoxazolin-2(3*H*)-one. J. Biol. Chem. 280: 21867–21881.

Baxter A, Mittler R and Suzuki N (2014) ROS as key players in plant stress signaling. J. Exp. Bot. 65: 1229–1240.

Behringer C, Bartsch K and Schaller A (2011) Safeners recruit multiple signalling pathways for the orchestrated induction of the cellular xenobiotic detoxification machinery in *Arabidopsis*. Plant Cell Environ. 34: 1970–1985.

Bonaventure G, Schuck S and Baldwin IT (2011) Revealing complexity and specificity in the activation of lipase-mediated oxylipin biosynthesis: a specific role of the *Nicotiana attenuata* GLA1 lipase in the activation of jasmonic acid biosynthesis in leaves and roots. Plant Cell Environ. 34: 1507–1520.

Brazier-Hicks M, Evans KM, Cunningham OD, Hodgson DRW, Steel P and Edwards R (2008) Catabolism of glutathione conjugates in *Arabidopsis thaliana*: role in metabolic reactivation of the herbicide safener fenclorim. J. Biol. Chem. 283: 21102–21112.

Breaux EJ, Hoobler MA, Patanella JE and Leyes GA (1989) Mechanisms of action of thiazole safeners. pp. 163–175. *In*: Hatzios KK and Hoagland RE (eds.). Crop Safeners for Herbicides. Development, Uses, and Mechanisms of Action. San Diego: Academic Press.

Christeller JT and Galis I (2014) alpha-Linolenic acid concentration and not wounding *per se* is the key regulator of octadecanoid (oxylipin) pathway activity in rice (*Oryza sativa* L.) leaves. Plant Physiol. Biochem. 83: 117–125.

Christensen SA, Nemchenko A, Park Y-S, Borrego E, Huang P-C, Schmelz EA, Kunze S, Feussner I, Yalpani N, Meeley R and Kolomiets MV (2014) The novel monocot-specific 9-lipoxygenase ZmLOX12 is required to mount an effective jasmonate-mediated defense against *Fusarium verticillioides* in maize. Mol. Plant Microbe. Interact. 27: 1263–1276.

Cigan AM, Unger-Wallace E, inventors; Pioneer Hi-Bred International, Inc, assignee (2015) April 9. Inducible promoter sequences for regulated expression and methods of use. US patent application 20150101077 A1.

Cole DJ (1994) Detoxification and activation of agrochemicals in plants. Pestic. Sci. 42: 209–222.

Cummins I, Dixon DP, Freitag-Pohl S, Skipsey M and Edwards R (2011) Multiple roles for plant glutathione transferases in xenobiotic detoxification. Drug Metab. Rev. 43: 266–280.

Dave A and Graham IA (2012) Oxylipin signaling: a distinct role for the jasmonic acid precursor cis-(+)-12-oxo-phytodienoic acid (cis-OPDA). Front Plant Sci. 3: 42. DOI: 10.3389/fpls.2012.00042.

Davies J and Caseley JC (1999) Herbicide safeners: A review. Pestic. Sci. 55: 1043–1058.

DeRidder BP, Dixon DP, Beussman DJ, Edwards R and Goldsbrough PB (2002) Induction of glutathione *S*-transferases in *Arabidopsis* by herbicide safeners. Plant Physiol. 130: 1497–1505.

DeRidder BP and Goldsbrough PB (2006) Organ-specific expression of glutathione *S*-transferases and the efficacy of herbicide safeners in *Arabidopsis*. Plant Physiol. 140: 167–175.

De Veylder L, Van Montagu M and Inze D (1997) Herbicide safener-inducible gene expression in *Arabidopsis thaliana*. Plant Cell Physiol. 38: 568–577.

Dixon DP and Edwards R (2009) Selective binding of glutathione conjugates of fatty acid derivatives by plant glutathione transferases. J. Biol. Chem. 284: 21249–21256.

Dixon DP, Skipsey M and Edwards R (2010) Roles for glutathione transferases in plant secondary metabolism. Phytochemistry 71: 338–350.

Duke SO (2012) Why have no new herbicide modes of action appeared in recent years? Pest Manag. Sci. 68: 505–512.

Dueckershoff K, Mueller S, Mueller MJ and Reinders J (2008) Impact of cyclopentenone-oxylipins on the proteome of *Arabidopsis thaliana*. Biochim. Biophys. Acta 1784: 1975–1985.

Ezra G, Rusness DG, Lamoureux GL and Stephenson GR (1985) The effect of CDAA (*N,N*-diallyl-2-chloroacetamide) pretreatments on subsequent CDAA injury to corn (*Zea mays* L.). Pestic. Biochem. Physiol. 23: 108–115.

Farago S, Brunhold C and Kreuz K (1994) Herbicide safeners and glutathione metabolism. Physiol. Plant 91: 537–542.

Farmer EE and Davoine C (2007) Reactive electrophile species. Curr. Opin. Plant Biol. 10: 380–386.

Ferhatoglu Y, Avdiushko S and Barrett M (2005) The basis for the safening of clomazone by phorate insecticide in cotton and inhibitors of cytochrome P450s. Pestic. Biochem. Physiol. 81: 59–70.

Frova C (2003) The plant glutathione transferase gene family: genomic structure, functions, expression and evolution. Physiol. Plant 119: 469–479.

Frova C (2006) Glutathione transferases in the genomics era: new insights and perspectives. Biomol. Eng. 23: 149–169.

Fuerst EP and Gronwald JW (1986) Induction of rapid metabolism of metolachlor in sorghum (*Sorghum bicolor*) shoots by CGA-92194 and other antidotes. Weed Sci. 34: 354–361.

Grebner W, Stingl NE, Oenel A, Mueller MJ and Berger S (2013) Lipoxygenase6-dependent oxylipin synthesis in roots is required for abiotic and biotic stress resistance of *Arabidopsis*. Plant Physiol. 161: 2159–2170.

Green JM (1989) Herbicide antagonism at the whole plant level. Weed Technol. 3: 217–226.

Green JM (1991) Maximizing herbicide efficiency with mixtures and expert systems. Weed Technol. 5: 894–897.

Green JM (2014) Current state of herbicides in herbicide-resistant crops. Pest Manag. Sci. 70: 1352–1357.

Green JM and Amuti KS (1993) Maximizing the performance of antagonistic mixtures. Acta Phytopath et Entomol. Hungar. 28: 469–480.

Hansch C and Muir RM (1950) The ortho effect in plant growth-regulators. Plant Physiol. 25: 389–393.

Hatzios KK (1983) Herbicide antidotes: Development, chemistry and mode of action. Adv. Agron. 36: 265–316.

Hatzios KK (1991) An overview of the mechanisms of action of herbicide safeners. Z Naturforsch J. Biosci. 46c: 819–827.

Hatzios KK and Burgos N (2004) Metabolism-based herbicide resistance: regulation by safeners. Weed Sci. 52: 454–467.

Hatzios KK and Hoagland RE (eds.) (1989) Crop Safeners for Herbicides. Development, Uses, and Mechanisms of Action. San Diego: Academic Press. 414 p.

Hershey HP and Stoner TD (1991) Isolation and characterization of cDNA clones for RNA species induced by substituted benzenesulfonamides in corn. Plant Mol. Biol. 17: 679–690.

Hoffmann OL (1953) Inhibition of auxin effects by 2,4,6-trichlorophenoxyacetic acid. Plant Physiol. 28: 622–628.

Holt DC, Lay VJ, Clarke ED, Dinsmore A, Jepson I, Bright SWJ and Greenland AJ (1995) Characterization of the safener-induced glutathione *S*-transferase isoform II from maize. Planta 196: 295–302.

Irzyk GP and Fuerst EP (1997) Characterization and induction of maize glutathione *S*-transferases involved in herbicide detoxification. pp. 155–170. *In*: Hatzios KK (ed.). Regulation of Enzymatic Systems Detoxifying Xenobiotics in Plants. Dordrecht, The Netherlands: Kluwer Academic Publishers.

Jablonkai I (2013) Herbicide safeners: effective tools to improve herbicide selectivity. pp. 589–620. *In*: Price AJ and Kelton JA (eds.). Herbicides—Current Research and Case Studies in Use. InTech: Rijeka, Croatia.

Jachetta JJ and Radosevich SR (1981) Enhanced degradation of atrazine by corn (*Zea mays*). Weed Sci. 29: 37–44.

Javelle M, Vernoud V, Rogowsky PM and Ingram GC (2011) Epidermis: the formation and functions of a fundamental plant tissue. New Phytol. 189: 17–39.

Jeschke P (2016) Propesticides and their use as agrochemicals. Pest Manag. Sci. 72: 210–225.

Keifer DW, inventor; FMC Corporation, assignee (2005) February 15. Method for safening crops from the phytotoxic effects of herbicides by use of phosphorated esters. US patent 6,855,667.

Komives T and Hatzios KK (1991) Chemistry and structure-activity relationships of herbicide safeners. Z Naturforsch J. Biosci. 46c: 798–804.

Kraehmer H, Laber B, Rosinger C and Schulz A (2014) Herbicides as weed control agents: state of the art. I. Weed control research and safener technology: the path to modern agriculture. Plant Physiol. 166: 1119–1131.

Kreuz K, Gaudin J and Ebert E (1989) Effects of the safeners CGA 154281, oxabetrinil and fenclorim on uptake and degradation of metolachlor in corn (*Zea mays* L.) seedlings. Weed Res. 29: 399–405.

Kreuz K, Tommasini R and Martinoia E (1996) Old enzymes for a new job. Herbicide detoxification in plants. Plant Physiol. 111: 349–353.

Lay M-M and Casida JE (1976) Dichloracetamide antidotes enhance thiocarbamate sulfoxide detoxification by elevating corn root glutathione content and glutathione S-transferase activity. Pestic. Biochem. Physiol. 6: 442–446.

Lay M-M and Casida JE (1978) Involvement of glutathione content and glutathione S-transferase in the action of dichloroacetamide antidotes for thiocarbamate herbicides. pp. 151–160. *In*: Pallos FM and Casida JE (eds.). Chemistry and Action of Herbicide Antidotes. New York: Academic Press.

Lay MM, Hubbell JP and Casida JE (1975) Dichloroacetamide antidotes for thiocarbamate herbicides: mode of action. Science 189: 287–288.

Liu J, Brazier-Hicks M and Edwards R (2009) A kinetic model for the metabolism of the herbicide safener fenclorim in *Arabidopsis thaliana*. Biophys. Chem. 143: 85–94.

Loeffler C, Berger S, Guy A, Durand T, Bringmann G, Dreyer M, von Rad U, Durner J and Mueller MJ (2005) B_1-phytoprostanes trigger plant defense and detoxification responses. Plant Physiol. 137: 328–340.

Mashiyama ST, Malabanan MM, Akiva E, Bhosle R, Branch MC et al. (2014) Large-scale determination of sequence, structure, and function relationships in cytosolic glutathione transferases across the biosphere. PLOS Biol. 12(4): e1001843.

Mosblech A, Feussner I and Heilmann I (2009) Oxylipins: structurally diverse metabolites from fatty acid oxidation. Plant Physiol. Biochem. 47: 511–517.

Mueller MJ (2004) Archetype signals in plants: the phytoprostanes. Curr. Opin. Plant Biol. 7: 441–448.

Mueller MJ and Berger S (2009) Reactive electrophilic oxylipins: pattern recognition and signalling. Phytochemistry 70: 1511–1521.

Mueller MJ, Mène-Saffrané L, Grun C, Karg K and Farmer EE (2006) Oxylipin analysis methods. Plant J. 45: 472–489.

Mueller S, Hilbert B, Dueckershoff K, Roitsch T, Krischke M, Mueller MJ and Berger S (2008) General detoxification and stress responses are mediated by oxidized lipids through TGA transcription factors in *Arabidopsis*. Plant Cell 20: 768–785.

Okazaki Y and Saito K (2014) Roles of lipids as signaling molecules and mitigators during stress response in plants. Plant J. 79: 584–596.

Padidam M (2003) Chemically regulated gene expression in plants. Curr. Opin. Plant Biol. 6: 169–177.

Pallett KE, Cramp SM, Little JP, Veerasekaran P, Crudace AJ and Slater AE (2001) Isoxaflutole: the background to its discovery and the basis of its herbicidal properties. Pest Manag. Sci. 57: 133–142.

Parker C (1983) Herbicide antidotes—a review. Pestic. Sci. 14: 40–48.

Philbrook B and Santel HJ (2008) A new formulation of isoxaflutole for pre-emergence weed control in corn (*Zea mays*). Weed Sci. Soc. Amer. Abstr. 48: 116.

Ramel F, Sulmon C, Serra A-A, Gouesbet G and Couee I (2012) Xenobiotic sensing and signaling in higher plants. J. Exp. Bot. 63: 3999–4014.

Riechers DE, Zhang Q, Xu FX and Vaughn KC (2003) Tissue-specific expression and localization of safener-induced glutathione S-transferase proteins in *Triticum tauschii*. Planta 217: 831–840.

Riechers DE, Vaughn KC and Molin WT (2005) The role of plant glutathione S-transferases in herbicide metabolism. pp. 216–232. *In*: Clark JM and Ohkawa H (eds.). Environmental

Fate and Safety Management of Agrochemicals. ACS Symposium Series 899. Washington DC: American Chemical Society.

Riechers DE, Kreuz K and Zhang Q (2010) Detoxification without intoxication: herbicide safeners activate plant defense gene expression. Plant Physiol. 153: 3–13.

Rishi A, Muni S, Kapur V, Nelson ND and Goyal A (2004) Identification and analysis of safener-inducible expressed sequence tags in *Populus* using a cDNA microarray. Planta 220: 296–306.

Rosinger CH (2015) A product portfolio for selective weed control in corn: flexible solutions based on various herbicide and safener assets. Weed Sci. Soc. Amer. Abstr. 55: 270.

Rosinger C, Bartsch K and Schulte W (2012) Safeners for herbicides. pp. 371–398. *In*: Krämer W, Schirmer U, Jeschke P and Witschel M (eds.). Modern Crop Protection Compounds. Vol. 1. Weinheim, Germany: Wiley-VCH.

Sari-Gorla M, Ferrario S, Rossini L, Frova C and Villa M (1993) Developmental expression of glutathione-S-transferase in maize and its possible connection with herbicide tolerance. Euphytica 67: 221–230.

Savchenko T, Kolla VA, Wang C-Q, Nasafi Z, Hicks DR, Phadungchob B, Chehab WE, Brandizzi F, Froehlich J and Dehesh K (2014) Functional convergence of oxylipin and abscisic acid pathways controls stomatal closure in response to drought. Plant Physiol. 164: 1151–1160.

Schaller A and Stintzi A (2009) Enzymes in jasmonate biosynthesis: structure, function, regulation. Phytochemistry 70: 1532–1538.

Schuck S, Kallenbach M, Baldwin IT and Bonaventure G (2014) The *Nicotiana attenuata* GLA1 lipase controls the accumulation of *Phytophthora parasitica*-induced oxylipins and defensive secondary metabolites. Plant Cell Environ. 37: 1703–1715.

Schulte W and Köcher H (2009) Tembotrione and combination partner isoxadifen-ethyl—mode of herbicidal action. Bayer. Crop Sci. J. 62(1): 35–52.

Skipsey M, Knight KM, Brazier-Hicks M, Dixon DP, Steel PG and Edwards R (2011) Xenobiotic responsiveness of *Arabidopsis thaliana* to a chemical series derived from a herbicide safener. J. Biol. Chem. 286: 32268–32276.

Stintzi A, Weber H, Reymond P, Browse J and Farmer EE (2001) Plant defense in the absence of jasmonic acid: the role of cyclopentenones. Proc. Natl. Acad. Sci. USA 98: 12837–12842.

Stotz HU, Mueller S, Zoeller M, Mueller MJ and Berger S (2013) TGA transcription factors and jasmonate-independent COI1 signalling regulate specific plant responses to reactive oxylipins. J. Exp. Bot. 64: 963–975.

Taki N, Sasaki-Sekimoto Y, Obayashi T, Kikuta A, Kobayashi K, Ainai T, Yagi K, Sakurai N, Suzuki H, Masuda T, Takamiya K-I, Shibata D, Kobayashi Y and Ohta H (2005) 12-Oxophytodienoic acid triggers expression of a distinct set of genes and plays a role in wound-induced gene expression in *Arabidopsis*. Plant Physiol. 139: 1268–1283.

Tam AC, Behki RM and Khan SU (1988) Effect of dietholate (R-33865) on the degradation of thiocarbamate herbicide by an EPTC-degrading bacterium. J. Agric. Food Chem. 36: 654–657.

Taylor VL, Cummins I, Brazier-Hicks M and Edwards R (2013) Protective responses induced by herbicide safeners in wheat. Environ. Exp. Bot. 88: 93–99.

Theodoulou FL, Clark IM, He X-L, Pallett KE, Cole DJ and Hallahan DL (2003) Co-induction of glutathione *S*-transferases and multidrug resistance associated protein by xenobiotics in wheat. Pest Manag. Sci. 59: 202–214.

Vicente J, Cascon T, Vicedo B, Garcia-Agustin P, Hamberg M and Castresana C (2012) Role of 9-lipoxygenase and α-dioxygenase oxylipin pathways as modulators of local and systemic defense. Mol. Plant 5: 914–928.

Ward ER, Ryals JA and Miflin BJ (1993) Chemical regulation of transgene expression in plants. Plant Mol. Biol. 22: 361–366.

Yan Y, Christensen S, Isakeit T, Engelberth J, Meeley R, Hayward A, Emery RJN and Kolomiets MV (2012) Disruption of OPR7 and OPR8 reveals the versatile functions of jasmonic acid in maize development and defense. Plant Cell 24: 1420–1436.

Yelverton FH, Worsham AD and Peedin GF (1993) Use of activated carbon to reduce phytotoxicity of imazaquin to tobacco (*Nicotiana tabacum*). Weed Technol. 7: 663–669.

Yu Q and Powles S (2014) Metabolism-based herbicide resistance and cross-resistance in crop weeds: A threat to herbicide sustainability and global crop production. Plant Physiol. 166: 1106–1118.

Zhang Q and Riechers DE (2004) Proteomic characterization of herbicide safener-induced proteins in the coleoptile of *Triticum tauschii* seedlings. Proteomics 4: 2058–2071.

Zhang Q, Xu FX, Lambert KN and Riechers DE (2007) Safeners coordinately induce the expression of multiple proteins and MRP transcripts involved in herbicide metabolism and detoxification in *Triticum tauschii* seedling tissues. Proteomics 7: 1261–1278.

Zhou G, Ren N, Qi J, Lu J, Xiang C, Ju H, Cheng J and Lou Y (2014) The 9-lipoxygenase *Osr9-LOX1* interacts with the 13-lipoxygenase-mediated pathway to regulate resistance to chewing and piercing-sucking herbivores in rice. Physiol. Plant 152: 59–69.

Chapter 8

Recent Advances in Deciphering Metabolic Herbicide Resistance Mechanisms

Vijay K. Nandula

Introduction

The next and most important phase after the confirmation of herbicide resistance in a weed population is the deciphering of the underlying resistance mechanism(s). The mechanism of resistance to an herbicide in a weed population can greatly determine the effectiveness of resistance management strategies. For example, a target site mutation could endow cross resistance to herbicides with similar mechanism of action, or metabolic resistance can bestow ability to withstand herbicides across more than one mechanism of action. In general, five modes of herbicide resistance have been identified in weeds: (1) altered target site due to a mutation at the site of herbicide action resulting in complete or partial lack of inhibition; (2) metabolic deactivation, whereby the herbicide active ingredient is transformed to nonphytotoxic metabolites; (3) reduced absorption and/or translocation that results in restricted movement of lethal levels of herbicide to point/site of action; (4) sequestration/compartmentation by which a herbicide is immobilized away from the site of action in cell organelles such as vacuoles or cell walls; and (5) gene amplification/over-expression of the target site with consequent dilution of the herbicide in relation to the target site (Nandula 2010). Among these, mutation at target site and gene

Crop Production Systems Research Unit, Agricultural Research Service, United States Department of Agriculture, Stoneville, MS 38776.
Email: vijay.nandula@ars.usda.gov

amplification are target site based, and metabolic deactivation, differential absorption/translocation, and sequestration are classified as non-target site-based resistances (NTSR).

The majority of research studies investigating differential absorption, translocation (research procedures of which have been recently summarized by Nandula and Vencill 2015), and metabolism are based on availability and application of ^{14}C-labeled herbicides. Shaner (2009) elegantly described differential translocation of glyphosate as a resistance mechanism. Herbicide sequestration, especially, sequestration of glyphosate, as a resistance mechanism has been investigated and reported on extensively by Ge et al. (2010, 2011, 2012) and reviewed by Sammons and Gaines (2014). This article aims to summarize current understanding of metabolic resistance in weeds by providing a history of related research, reporting recent advances, and identifying future research opportunities.

Herbicide Tolerance

The Weed Science Society of America (WSSA) defines herbicide tolerance as "the inherent ability of a species to survive and reproduce after herbicide treatment" (WSSA 1998). This implies that there was no selection or genetic manipulation to make the plant tolerant; it is naturally tolerant. Differential herbicide metabolism is one of the most common mechanisms by which crop plants tolerate phytotoxic action of herbicides, while wild type/ susceptible weeds are controlled. Metabolism of herbicides usually occurs in three phases (Kreuz et al. 1996; Van Eerd et al. 2003): a conversion of the herbicide molecule into a more hydrophilic metabolite (phase 1), followed by conjugation to biomolecules such as glutathione/sugar (phase 2), and further conjugation/breakup/oxidation reactions followed by transport to vacuoles or cell walls where additional breakdown occurs (phase 3) (Delye 2013).

Safeners

In general, safeners (focus of a different chapter in this book) are chemicals applied in combination with herbicides to provide tolerance to grass crops such as wheat, rice, corn, and sorghum against certain thiocarbamate, chloroacetamide, sulfonylurea, and aryoxyphenoxypropionate herbicides that are applied preemergence or postemergence. Safeners enhance herbicide detoxification in 'safened' plants. Safening agents activate/ catalyze cofactors such as glutathione and enzyme systems such as cytochrome P450 monoxygenases, glutathione S-transferases, and glycosyl transferases (Hatzios and Burgos 2004). The safener-mediated induction of herbicide-detoxifying enzymes appears to be part of a general stress response (Delye 2013; Hatzios and Burgos 2004). These enzymes deactivate herbicide molecules by modifying side chains, which then are conjugated to biochemical moieties such as sugar and amino acid residues. Some of these conjugates are further deposited in vacuoles and cell walls.

Cytochrome P450 Monoxygenase

Cytochrome P450 monoxygenases (CYP, E.C. 1.14.13.X) are oxidative enzymes that have the most important role in Phase 1 of herbicide metabolism (Barrett 2000). CYP often catalyze monooxygenase reactions, usually resulting in hydroxylation, according to the following reaction: RH + O_2 + NAD(P)H + H^+ -> ROH + H_2O + NAD(P)$^+$ (Van Eerd et al. 2003). CYP can be divided into three classes (Van Eerd et al. 2003). Class I CYP require flavin adenine dinucleotide (FAD), or flavin mononucleotide (FMN), or reduced nicotinamide adenine dinucleotide phosphate (NADPH), and are usually microsomal membrane-bound proteins in plants and filamentous fungi. Class II CYP, while similar to class I, exist only in bacterial and animal mitochondria. Class III CYP occur in plant plastids and do not require auxillary redox partners (Van Eerd et al. 2003).

A comprehensive list of herbicides subjected to *in vitro* CYP-mediated metabolic reactions was compiled by Siminszky (2006). Described below are selected discoveries on CYP enzymes in selected crops including original reports as well as recently documented cases.

Arabidopsis. Three CYP enzymes, CYP76C1, CYP76C2, and CYP76C4 of CYP76C subfamily specific to Brassicaceae, metabolized herbicides belonging to the class of phenylurea in *Arabidopsis thaliana* (L.) (Hofer et al. 2014). These CYPs also metabolized natural monoterpenols.

Corn. Polge and Barrett (1995) provided evidence of the occurrence of a CYP involved in the metabolism of chlorimuron-ethyl in corn microsomal preparations.

Lupin. Tolerance to metribuzin in mutants of narrow-leafed lupin (*Lupinus angustifolius* L.) was reversed after treatment with CYP inhibitors omethoate, malathion, and phorate (Pan et al. 2012).

Rice. A CYP-mediated O-demethylation of bensulfuron-methyl (BSM) played an important role in the metabolism of BSM by rice (*Oryza sativa* L. cv. Lemont) seedlings (Deng and Hatzios 2003). A novel CYP, *CYP81A6*, encoded by *Bel*, a gene found in wild type rice, provided resistance to two herbicide classes, bentazon (PS II inhibitor) and sulfonylureas, in two mutant male sterile hybrid rice parent lines (Pan et al. 2006). A novel rice CYP, *CYP72A31*, was involved in bispyribac-sodium (BS) tolerance (Saika et al. 2014). BS tolerance was correlated with *CYP72A31* mRNA levels in transgenic plants of rice and *A. thaliana*. Moreover, *Arabidopsis* overexpressing *CYP72A31* showed tolerance to BSM, which belongs to a different class of ALS-inhibiting herbicides. On the other hand, CYP81A6, which has been reported to confer BSM tolerance, was barely involved, if at all, in BS tolerance, suggesting that the *CYP72A31* enzyme has different herbicide specificities compared to *CYP81A6*.

Soybean. The gene product of a CYP cDNA (*CYP71A10*) from soybean, expressed in yeast, specifically catalyzed the metabolism of phenylurea herbicides, converting four herbicides of this class (fluometuron, linuron, chlortoluron, and diuron) into more polar compounds (Siminszky et al. 1999). Analyses of the metabolites suggested that the *CYP71A10* encoded enzyme functions primarily as an *N*-demethylase with regard to fluometuron, linuron, and diuron, and as a ring-methyl hydroxylase when chlortoluron is the substrate. *In vivo* assays using excised leaves demonstrated that all four herbicides were more readily metabolized in *CYP71A10*-transformed tobacco compared with control plants. For linuron and chlortoluron, *CYP71A10*-mediated herbicide metabolism resulted in significantly enhanced tolerance to these compounds in the transgenic plants (Siminszky et al. 1999).

Wheat. Wheat (*Triticum aestivum* L. cv Etoile de Choisy) microsomes catalyzed the CYP-dependent oxidation of the herbicide diclofop to three hydroxy-diclofop isomers (Zimmerlin and Durst 1992).

Glutathione *S*-Transferase

Glutathione *S*-transferases (GST, E.C. 2.5.1.18) are a broadly present, multifunctional family of enzymes that catalyze the conjugation of glutathione to a variety of substrates (Marrs 1996), including herbicides. Glutathione is a tripeptide of glutamate-cysteine-glycine present in the cytosol of several organisms including plants. It acts as an antioxidant by minimizing membrane damage from reactive oxygen species (free radicals, peroxide, etc.) and itself is oxidized to glutathione disulfide in the process. GST isozymes are cytosolic, in general, and exist as homo- and heterodimers with a subunit molecular mass of 25 kDA (Droog 1997). Glutathione-*S*-conjugate uptake into the plant vacuole is mediated by a specific ATPase which is remarkably similar to the glutathione-*S*-conjugate export pumps in the canalicular membrane of mammalian liver (Martinoia et al. 1993).

Described below are selected discoveries on GST enzymes in selected crops including original reports as well as recently documented cases.

Arabidopsis. The three-dimensional structure of GST from *A. thaliana* indicated the lack of tyrosine in its active site, as opposed to mammalian GSTs, which share a conserved catalytic tyrosine residue (Reinemer et al.

1996). A transporter responsible for the removal of glutathione S-conjugates from the cytosol, a specific Mg21-ATPase, is encoded by the *AtMRP1* gene of *A. thaliana* (Lu et al. 1997). The sequence of *AtMRP1* and the transport capabilities of membranes prepared from yeast cells transformed with plasmid-borne *AtMRP1* demonstrate that this gene encodes an ATP-binding cassette transporter competent in the transport of glutathione S-conjugates of xenobiotics and endogenous substances, including herbicides and anthocyanins. At*GSTU19*, a tau class GST, when expressed in *Escherichia coli* was highly active towards chloroacetanilide herbicides (DeRidder et al. 2002).

Corn. The first report of GST activity in plants was on corn (Frear and Swanson 1970). Glutathione conjugation of atrazine, first example of biotransformation of a pesticide in plants, is the primary mechanism of tolerance of corn to the herbicide (Shimabukuro et al. 1970). GST-III was isolated from *Z. mays* var. *mutin* and was cloned, sequenced, and its structure determined by x-ray crystallography (Neuefeind et al. 1997). The enzyme forms a GST-typical dimer with one subunit consisting of 220 residues. Each subunit is formed of two distinct domains, an N-terminal domain consisting of a β-sheet flanked by two helices, and a C-terminal domain, entirely helical. The dimeric molecule is globular with a large cleft between the two subunits.

Rice. A GST from rice showed 44–66% similarity to the sequences of the class phi GSTs from *A. thaliana* and corn (Cho and Kong 2005). The isolated gene product, OsGSTF3-3, displayed high activity toward 1-chloro-2,4-dinitrobenzene, a general GST substrate and also had high activities towards the acetanilide herbicides, alachlor, and metolachlor.

Soybean. GSTs in soybean (*Gm*GSTs) involved in herbicide detoxification in cell suspension cultures were purified (Andrews et al. 2005). With respect to herbicide detoxification, two *Gm*GSTU2-related polypeptides dominated the activity toward the chloroacetanilide acetochlor, while an unclassified subunit was uniquely associated with the detoxification of diphenyl ethers (acifluorfen, fomesafen). The inducibility of the diverent GST subunits was determined in soybean plants exposed to photobleaching diphenyl ethers and the safeners naphthalic anhydride and dichlormid. *Gm*GSTU3, a *Gm*GSTU1-like polypeptide, and thiol (homoglutathione) content were induced by all chemical treatments, while two uncharacterized subunits were only induced in plants showing photobleaching.

Glycosyl Transferase

Glycosyl transferases (GTs, E.C. 2.4) comprise of a large gene family in which proteins conjugate a sugar molecule to a wide range of lipophilic small molecule acceptors including herbicides (Bowles et al. 2005). The conjugation reactions enable GTs to diversify the secondary metabolites via sugar attachments, to maintain cell homeostasis by quickly and precisely controlling plant hormone concentration, as well as to detoxify herbicides by adding sugars onto molecules (Yuan et al. 2006). Glycosyl transferases exist as a gene superfamily with diverse members. They are found in all kingdoms and can be classified into 78 subfamilies. Two soybean GTs were shown to glycosylate the primary major bentazone metabolite, 6-hydroxybentazone (Leah et al. 1992). Other GTs with activity towards herbicides such as 2,4,5-trichlorophenol have since been cloned and characterized (Brazier-Hicks and Edwards 2005; Loutre et al. 2003).

Metabolic Resistance in Weeds

Metabolic resistance to herbicides in weed species has been studied over the past several decades, but not as extensively or in depth as the research on target site-based resistance. An obvious reason is the difficulty in unraveling the complicated physiological and biochemical processes resulting in the metabolic resistance mechanism. In North (US and Canada) and South American (Brazil and Argentina) crop production fields, the relative ease of cultivation of glyphosate resistant crops drastically minimized adverse economic impact from the evolution of herbicide resistant weeds, until the emergence of glyphosate resistant Palmer amaranth (*Amaranthus palmeri* S. Wats.) (Culpepper et al. 2005). Meanwhile, progress has been made over the past decade, more so in the last two to three years, in understanding metabolic resistance in grass weed species (detailed in following sections) through techniques such as RNA-seq and next generation sequencing.

Blackgrass. Metabolic herbicide resistance, involving CYP, has been identified in several European blackgrass populations (Cocker et al. 1999; De Prado and Franco 2004; Letouzé and Gasquez 2001, 2003; Menendez and De Prado 1997). When a black-grass GST, *AmGSTF1* was expressed in *A. thaliana*, the transgenic plants acquired resistance to multiple herbicides and showed similar changes in their secondary, xenobiotic, and antioxidant metabolism to those determined in multiple herbicide resistant weeds (Cummins et al. 2013). Transcriptome array experiments showed that these changes in biochemistry were not due to changes in gene expression. On the contrary, *AmGSTF1* exerted a direct regulatory control on metabolism that led to an accumulation of protective flavonoids. In addition, a multiple drug resistance inhibiting pharmacophore, 4-chloro-7-nitro-benzoxadiazole, with applications in human cancer tumor research, was active on *Am*GSTF1 and helped restore herbicide susceptibility in multiple herbicide resistant blackgrass. Role of an induced GT in multiple herbicide resistant blackgrass (Brazier et al. 2002) was the first such evidence in weeds.

Rigid ryegrass. Resistance to diclofop in a rigid ryegrass population, selected by recurrent treatment with low doses of the herbicide, was due to rapid metabolism of diclofop acid (likely brought about by CYP) resulting in a 2.6-fold less of the phytotoxic acid compared to a susceptible population (Yu et al. 2013). The major polar metabolites of diclofop acid were similar to that of wheat, a crop naturally tolerant to diclofop. Pre-treatment with 2,4-D, a known CYP inducer, caused up to 10-fold change in LD_{50} and GR_{50} in dose–response to subsequently applied diclofop-methyl in a herbicide susceptible rigid ryegrass population (Han et al. 2013). Metabolism of diclofop acid, following of de-esterification of diclofop-methyl, to non phytotoxic metabolites was 1.8-fold faster in 2,4-D pre-treated plants than on untreated plants. Also, 2,4-D pre-treatment induced cross-protection against the ALS-inhibiting herbicide chlorsulfuron. RNA-Seq transcriptome analysis of diclofop-resistant rigid ryegrass identified four contigs, two CYP, a nitronate monoxygenase (NMO), and a GT, consistently highly expressed in nine field-evolved metabolic resistant populations (Gaines et al. 2014). These four contigs were strongly associated with the resistance phenotype and were major candidates for contributing to metabolic diclofop resistance. More than two-thirds of 33 diclofop resistant populations of rigid ryegrass exhibited both metabolic resistance and target-site ACCase mutations (Han et al. 2015). Duhoux et al. (2015) confirmed four contigs comprising of two CYP enzymes, one GT, and one GST in ryegrass plants resistant to pyroxsulam.

Echinochloa **spp.** The addition of CYP inhibitors, piperonyl butoxide and malathion, severely increased the sensitivity of a resistant watergrass [*Echinochloa phyllopogon* (Stapf) Koso-Pol.] accession from California to BS, suggesting that metabolic degradation of the herbicide is the primary mechanism (Fischer et al. 2000). Initiation of CYP activity was implicated as the mechanism of multiple resistances in a resistant watergrass biotype from California after greater CYP induction by BS, fenoxaprop-ethyl, and thiobencard in the resistant biotype compared to a susceptible biotype (Yun et al. 2005). Similar CYP-induced multiple resistances endowed cross resistance to clomazone in late watergrass (Yasuor et al. 2010). Clomazone is a proherbicide that must be metabolized to 5-ketoclomazone, which is the active compound. Resistant plants accumulated 6- to 12-fold more of a nonphytotoxic metabolite (containing a monohydroxylated isoxazolidinone ring) than susceptible plants, while susceptible plants accumulated 2.5-fold more of the phytotoxic metabolite of clomazone, 5-ketoclomazone. A late watergrass biotype, resistant to fenoxaprop, metabolized the herbicide to nonphytotoxic polar metabolites and phytotoxic fenoxaprop acid 2-fold more and 5-fold less, respectively, compared to a susceptible biotype (Bakkali et al. 2007). In addition, the resistant biotype exhibited higher rate of glutathione conjugation. Resistant *E. phyllopogan* plants metabolized

BSM via *O*-demethylation more rapidly than susceptible plants (Iwakami et al. 2014). Two CYP genes belonging to the *CYP81A* subfamily, *CYP81A12* and *CYP81A21*, were more abundantly transcribed in the resistant plants compared with susceptible plants. Transgenic *A. thaliana*, expressing either of the above two genes, survived BSM or penoxsulam in media, but not wild type plants. Proteins of *CYP81A12* and *CYP81A21*, produced heterologously in yeast (*Saccharomyces cerevisiae*), metabolized bensulfuron-methyl by *O*-demethylation.

A propanil resistant barnyardgrass [*E. crus-galli* L. Beauv.] population metabolized the herbicide to 3,4-dichloroaniline, but not the susceptible population (Carey et al. 1997). Two other polar metabolites found in resistant barnyardgrass were similar to those formed in rice. Further work detected elevated (2- to 4-fold) activity of aryl acylamidase in the propanil resistant barnyardgrass population compared to a susceptible population (Hirase and Hoagland 2006).

Bromus. *Bromus rigidus* (Roth) Lainz populations from Australia that were resistant to sulfosulfuron, an ALS-inhibiting herbicide, were reverted to being susceptible after treatment with the herbicide in combination with malathion, a CYP inhibitor (Owen et al. 2012).

***Avena* spp.** A diclofop-methyl resistant wild oat (*Avena* spp.) population, lacking a resistant acetylCoA carboxylase (ACCase), metabolized the parent methyl ester to the phytotoxic diclofop acid to a lesser extent than a susceptible population (Ahmad-Hamdani et al. 2013). In addition, there was an associated higher level (up to 1.7-fold) of nontoxic polar diclofop metabolites in the resistant plants relative to susceptible plants, indicating a non-target site based mechanism of enhanced rate of diclofop acid metabolism. Three other resistant populations had lower diclofop acid levels in addition to ACCase mutations.

Broadleaf weeds. Resistance to atrazine in velvetleaf populations from Wisconsin and Maryland was found to be due to metabolism of atrazine to glutathione, L-cysteine, and *N*-acetyl-L-cysteine conjugates, metabolites produced in the glutathione conjugation pathway (Anderson and Gronwald 1991; Gray et al. 1996; Gronwald et al. 1989). Resistance to atrazine, applied postemergence, in two waterhemp populations was due to increased levels of GST activity (Ma et al. 2013).

Future Research Direction

A three-step procedure was proposed, based on the use of the 'omics' (genomics, transcriptomics, proteomics or metabolomics), to decipher the genetic bases of NTSR (Delye 2013). Step 1 involves collection of weed genotypes, phenotype determination based on response to herbicide (resistant/susceptible), and production of genetically homogeneous plant material via controlled crosses. Step 2 is an omics-based approach to identify phenotype-related differences in gene expression to yield NTSR alleles. Step 3 pertains to the validation of NTSR candidate alleles with eventual development of DNA/protein/metabolite-based NTSR makers for NTSR diagnosis or NTSR evolutionary research.

Conclusions

The immediate and urgent challenge for weed scientists is to understand and characterize the reasons of NTSR, which includes metabolic resistance in order to sustain the limited herbicide portfolio and develop integrated weed management strategies (Délye 2013; Yu and Powles 2014). Metabolic resistance research in weeds has mostly been limited to grass species such as rigid ryegrass, blackgrass, and watergrass. However, dicot species such as tall waterhemp has developed resistance to multiple herbicide mechanisms of action by enhanced metabolic degradation (Ma et al. 2013). Thus, both grass and dicot species can develop metabolic herbicide resistance given the high initial frequency of genes responsible for imparting metabolic resistance. Herbicides must be used at full rates (Yu and Powles 2014) to minimize weed escapes, which if left uncontrolled can recharge weed seedbanks or evolve resistance.

LITERATURE CITED

Ahmad-Hamdani MS, Yu Q, Han H, Cawthray GR, Wang SF and Powles SB (2013) Herbicide resistance endowed by enhanced rates of herbicide metabolism in wild oat (*Avena* spp.). Weed Sci. 61: 55–62.

Anderson MP and Gronwald JW (1991) Atrazine resistance in a velvetleaf (*Abutilon theophrasti*) biotype due to enhanced glutathione *S*-transferase activity. Plant Physiol. 96: 104–109.

Andrews CJ, Cummins I, Skipsey M, Grundy NM, Jepson I, Townson J and Edwards R (2005) Purification and characterisation of a family of glutathione transferases with roles in herbicide detoxification in soybean (*Glycine max* L.); selective enhancement by herbicides and herbicide safeners. Pestic. Biochem. Physiol. 82: 205–219.

Bakkali Y, Ruiz-Santaella JP, Osuna MD, Wagner J, Fischer AJ and De Prado R (2007) Late watergrass (*Echinochloa phyllopogon*): Mechanisms involved in the resistance to fenoxaprop-*p*-ethyl. J. Agric. Food Chem. 55: 4052–4058.

Barrett M (2000) The role of cytochrome P450 enzymes in herbicide metabolism. pp. 25–37. *In*: Cobbs AH and Kirkwood RC (eds.). Herbicides and Their Mechanisms of Action. Sheffield, Great Britain: Sheffield Academic.

Bowles D, Isayenkova J, Lim EK and Poppenberger B (2005) Glycosyltransferases: managers of small molecules. Curr. Opin. Plant Biol. 8: 254–263.

Brazier M, Cole DJ and Edwards R (2002) *O*-glucosyltransferase activities toward phenolic natural products and xenobiotics in wheat and herbicide-resistant and herbicide-susceptible blackgrass (*Alopecurus myosuroides*). Phytochemistry 59: 149–156.

Brazier-Hicks M and Edwards R (2005) Functional importance of the family 1 glucosyltransferase *UGT72B1* in the metabolism of xenobiotics in *Arabidopsis thaliana*. Plant J. 42: 556–566.

Carey VF III, Hoagland RE and Talbert RE (1997) Resistance mechanism of propanil-resistant barnyardgrass: II. *In-vivo* metabolism of the propanil molecule. Pestic. Sci. 49: 333–338.

Cho H-Y and Kong K-H (2005) Molecular cloning, expression, and characterization of a phi-type glutathione *S*-transferase from *Oryza sativa*. Pestic. Biochem. Physiol. 83: 29–36.

Cocker KM, Moss SR and Coleman JOD (1999) Multiple mechanisms of resistance to fenoxaprop-P-ethyl in United Kingdom and other European populations of herbicide-resistant *Alopecurus myosuroides* (black grass). Pestic. Biochem. Physiol. 65: 169–180.

Culpepper AS, Grey TL, Vencill WK, Kichler JM, Webster TM, Brown SM, York AC, Davis JW and Hanna WW (2005) Glyphosate-resistant Palmer amaranth (*Amaranthus palmeri*) confirmed in Georgia. Weed Sci. 54: 620–626.

Cummins I, Wortley DJ, Sabbadin F, He Z, Coxona CR, Straker HE, Sellars JD, Knight K, Edwards L, Hughes D, Kaundun SS, Hutchings S-J, Steel PG and Edwards R (2013) Key role for a glutathione transferase in multiple-herbicide resistance in grass weeds. Proc. Natl. Acad. Sci. USA 110: 5812–5817.

Délye C (2013) Unravelling the genetic bases of non-target-site-based resistance (NTSR) to herbicides: a major challenge for weed science in the forthcoming decade. Pest Manag. Sci. 69: 176–187.

Deng F and Hatzios KK (2003) Characterization of cytochrome P450-mediated bensulfuron-methyl O-demethylation in rice. Pestic. Biochem. Physiol. 74: 102–115.

De Prado RA and Franco AR (2004) Cross-resistance and herbicide metabolism in grass weeds in Europe: biochemical and physiological aspects. Weed Sci. 52: 441–447.

DeRidder BP, Dixon DP, Beussman DJ, Edwards R and Goldsbrough PB (2002) Induction of glutathione S-transferases in *Arabidopsis* by herbicide safeners. Plant Physiol. 130: 1497–505.

Droog F (1997) Plant glutathione S-transferases, a tale of theta and tau. J. Plant Growth Regul. 16: 95–107.

Duhoux A, Carrère S, Gouzy J, Bonin L and Délye C (2015) RNA-Seq analysis of rye-grass transcriptomic response to an herbicide inhibiting acetolactate-synthase identifies transcripts linked to non-target-site-based resistance. Plant Mol. Biol. 87: 473–487.

Fischer AJ, Bayer DE, Carriere MD, Ateh CM and Yim K-O (2000) Mechanisms of resistance to bispyribac-sodium in an *Echinochloa phyllopogon* accession. Pestic. Biochem. Physiol. 68: 156–165.

Frear DS and Swanson HR (1970) Biosynthesis of *S*-(4-ethylamino-6-isopropylamino-2-*s*-triazino) glutathione: partial purification and properties of a glutathione *S*-transferase from corn. Phytochemistry 9: 2123–2132.

Gaines TA, Lorentz L, Figge A, Herrmann J, Maiwald F, Ott M-C, Han H, Busi R, Yu Q, Powles SB and Beffa R (2014) RNA-Seq transcriptome analysis to identify genes involved in metabolism-based diclofop resistance in *Lolium rigidum*. Plant J. 78: 865–876.

Ge X, d'Avignon DA, Ackerman JJH and Sammons RD (2010) Rapid vacuolar sequestration: the horseweed glyphosate resistance mechanism. Pest Manag. Sci. 66: 345–348.

Ge X, d'Avignon DA, Ackerman JJH, Duncan B, Spaur MB and Sammons RD (2011) Glyphosate-resistant horseweed made sensitive to glyphosate: low-temperature suppression of glyphosate vacuolar sequestration revealed by ^{31}P NMR. Pest Manag. Sci. 67: 1215–1221.

Ge X, d'Avignon DA, Ackerman JJ, Collavo A, Sattin M, Ostrander EL, Hall EL, Sammons RD and Preston C (2012) Vacuolar glyphosate-sequestration correlates with glyphosate resistance in ryegrass (*Lolium* spp.) from Australia, South America, and Europe: a ^{31}P NMR investigation. J. Agric. Food Chem. 60: 1243–1250.

Gray JA, Balke NE and Stoltenberg DE (1996) Increased glutathione conjugation of atrazine confers resistance in a Wisconsin velvetleaf (*Abutilon theophrasti*) biotype. Pestic. Biochem. Physiol. 55: 157–171.

Gronwald JW, Anderson RN and Yee C (1989) Atrazine resistance in velvetleaf (*Abutilon theophrasti*) due to enhanced atrazine detoxification. Pestic. Biochem. Physiol. 34: 149–163.

Han H, Yu Q, Cawthray GR and Powles SB (2013) Enhanced herbicide metabolism induced by 2,4-D in herbicide susceptible *Lolium rigidum* provides protection against diclofop-methyl. Pest Manag. Sci. 69: 996–1000.

Han H, Yu Q, Owen MJ, Cawthray GR and Powles SB (2015) Widespread occurrence of both metabolic and target-site herbicide resistance mechanisms in *Lolium rigidum* populations. Pest Manag. Sci. DOI 10.1002/ps.3995.

Hatzios KK and Burgos N (2004) Metabolism-based herbicide resistance: regulation by safeners. Weed Sci. 52: 454–467.

Hirase K and Hoagland RE (2006) Characterization of aryl acylamidase activity from propanil-resistant barnyardgrass (*Echinochloa crus-galli* [L.] Beauv.). Weed Biol. Manag. 6: 197–203.

Höfer R, Boachon B, Renault H, Gavira C, Miesch L, Iglesias J, Ginglinger J-F, Allouche L, Miesch M, Grec S, Larbat R and Werck-Reichhart D (2014) Dual function of the cytochrome P450 CYP76 family from *Arabidopsis thaliana* in the metabolism of monoterpenols and phenylurea herbicides. Plant Physiol. 166: 1149–1161.

Iwakami S, Endo M, Saika H, Okuno J, Nakamura N, Yokoyama M, Watanabe H, Toki S, Uchino A and Inamura T (2014) Cytochrome P450 CYP81A12 and CYP81A21 are associated with resistance to two acetolactate synthase inhibitors in *Echinochloa phyllopogon*. Plant Physiol. 165: 618–629.

Kreuz K, Tommasini R and Martinoia E (1996) Old enzymes for a new job. Plant Physiol. 111: 349–353.

Leah JM, Worrall TL and Cobb AH (1992) Isolation and characterization of 2 glucosyltransferases from *Glycine max* associated with bentazone metabolism. Pest Sci. 34: 81–87.

Letouzé A and Gasquez J (2001) Inheritance of fenoxaprop-P-ethyl resistance in a blackgrass (*Alopecurus myosuroides* Huds.) population. Theor. Appl. Genet. 103: 288–296.

Letouzé A and Gasquez J (2003) Enhanced activities of several herbicide degrading enzymes: a suggested mechanism responsible for multiple resistance in blackgrass (*Alopecurus myosuroides* Huds.). Agronomie 23: 601–608.

Loutre C, Dixon DP, Brazier M, Slater M, Cole DJ and Edwards R (2003) Isolation of a glucosyltransferase from *Arabidopsis thaliana* active in the metabolism of the persistent pollutant 3,4-dichloroaniline. Plant J. 34: 485–493.

Lu Y-P, Li Z-S and Rea PA (1997) *AtMRP1* gene of *Arabidopsis* encodes a glutathione *S*-conjugate pump: Isolation and functional definition of a plant ATP-binding cassette transporter gene. Proc. Natl. Acad. Sci. USA 94: 8243–8248.

Ma R, Kaundun SS, Tranel PJ, Riggins CW, McGinness DL, Hager AG, Hawkes T, McIndoe E and Riechers DE (2013) Distinct detoxification mechanisms confer resistance to mesotrione and atrazine in a population of waterhemp. Plant Physiol. 163: 363–377.

Marrs KA (1996) The functions and regulations of glutathione *S*-transferases in plants. Annu. Rev. Plant Physiol. Plant Mol. Biol. 47: 127–158.

Martinoia E, Grill E, Tommasini R, Kreuz K and Amrhein N (1993) ATP-dependent *S*-glutathione 'export' pump in the vacuolar membrane of plants. Nature 247–249.

Menendez J and De Prado R (1997) Diclofop-methyl cross-resistance in a chlorotoluron-resistant biotype of *Alopecurus myosuroides*. Pestic. Biochem. Physiol. 56: 123–133.

Nandula VK (2010) Herbicide resistance: definitions and concepts. pp. 35–43. *In*: Nandula VK (ed.). Glyphosate Resistance in Crops and Weeds: History, Development, and Management. Hoboken, NJ: John Wiley and Sons, Inc.

Nandula VK and Vencill WK (2015) Research methods in weed science: Herbicide absorption and translocation in plants using radioisotopes. Weed Sci. 63(Special Issue): 140–151.

Neuefeind T, Huber R, Reinemer P, Knäblein J, Prade L, Mann k and Bieseler B (1997) Cloning, sequencing, crystallization and x-ray structure of glutathione *S*-transferase-III from *Zea mays* var. *mutin*: a leading enzyme in detoxification of maize herbicides. Mol. Biol. 274: 577–587.

Owen MJ, Goggin DE and Powles SB (2012) Non-target-site-based resistance to ALS-inhibiting herbicides in six *Bromus rigidus* populations from Western Australian cropping fields. Pest Manag. Sci. 68: 1077–1082.

Pan G, Si P, Yu Q, Tu J and Powles S (2012) Non-target site mechanism of metribuzin tolerance in induced tolerant mutants of narrow-leafed lupin (*Lupinus angustifolius* L.). Crop Past Sci. 63: 452–458.

Pan G, Zhang X, Liu K, Zhang J, Wu X, Zhu J and Tu J (2006) Map-based cloning of a novel rice cytochrome P450 gene *CYP81A6* that confers resistance to two different classes of herbicides. Plant Mol. Biol. 61: 933–943.

Polge ND and Barrett M (1995) Characterization of cytochrome P450-mediated chlorimuron ethyl hydroxylation in maize microsomes. Pestic. Biochem. Physiol. 53: 193–204.

Reinemer P, Prade L, Hof P, Neuefeind T, Hube R, Zettl R, Palme K, Schell J, Koelln I, Bartunik HD and Bieseler B (1996) Three-dimensional structure of glutathione S-transferase from *Arabidopsis thaliana* at 2.2 Å resolution: structural characterization of herbicide-conjugating plant glutathione S-transferases and a novel site architecture. J. Mol. Biol. 255: 289–309.

Saika H, Horita J, Taguchi-Shiobara F, Nonaka S, Nishizawa-Yokoi A, Iwakami S, Hori K, Matsumoto T, Tanaka T, Itoh T, Yano M, Kaku K, Shimizu T and Toki S (2014) A novel rice cytochrome P450 gene, *CYP72A31*, confers tolerance to acetolactate synthase-inhibiting herbicides in rice and Arabidopsis. Plant Physiol. DOI:10.1104/pp.113.231266.

Sammons RD and Gaines TA (2014) Glyphosate resistance: state of knowledge. Pest Manag. Sci. 70: 1367–1377.

Shaner DL (2009) Role of translocation as a mechanism of resistance to glyphosate. Weed Sci. 57: 118–123.

Shimabukuro RH, Swanson HR and Walsh WC (1970) Glutathione conjugation: Atrazine detoxication mechanism in corn. Plant Physiol. 46: 103–107.

Siminszky B (2006) Plant cytochrome P450-mediated herbicide metabolism. Phytochem. Rev. 5: 445–458.

Siminszky B, Corbin FT, Ward ER, Fleischmann TJ and Dewey RE (1999) Expression of a soybean cytochrome P450 monooxygenase cDNA in yeast and tobacco enhances the metabolism of phenylurea herbicides. Proc. Natl. Acad. Sci. USA 96: 1750–1755.

Van Eerd LL, Hoagland RE, Zablotowicz RM and Hall JC (2003) Pesticide metabolism in plants and microorganisms. Weed Sci. 51: 472–495.

Weed Science Society of America (WSSA) (1998) Herbicide resistance and herbicide tolerance defined. Weed Technol. 12: 789.

Yasuor H, Zou W, Tolstikov VV, Tjeerdema RS and Fischer AJ (2010) Differential oxidative metabolism and 5-ketoclomazone accumulation are involved in *Echinochloa phyllopogon* resistance to clomazone. Plant Physiol. 153: 319–326.

Yu Q, Han H, Cawthray GR, Wang SF and Powles SB (2013) Enhanced rates of herbicide metabolism in low herbicide-dose selected resistant *Lolium rigidum*. Plant Cell Environ. 36: 818–27.

Yu Q and Powles S (2014) Metabolism-based herbicide resistance and cross-resistance in crop weeds: a threat to herbicides sustainability and global crop production. Plant Physiol. 166: 1106–1118.

Yuan JS, Tranel PJ and Stewart Jr CN (2006) Non-target-site herbicide resistance: a family business. Trends Plant Sci. 12: 6–13.

Yun M-S, Yogo Y, Miura R, Yamasue Y and Fischer AJ (2005) Cytochrome P-450 monooxygenase activity in herbicide-resistant and -susceptible late watergrass (*Echinochloa phyllopogon*). Pestic. Biochem. Physiol. 83: 107–114.

Zimmerlin A and Durst F (1992) Aryl hydroxylation of the herbicide diclofop by a wheat cytochrome P-450 monooxygenase. Plant Physiol. 100: 874–881.

Chapter 9

Recent Advances in Target Site Herbicide Resistance Mechanisms

Christopher Preston

Introduction

Types of Target Site Resistance Mechanisms that Occur in Weed Species under Selection from Herbicides

All herbicides cause plant death through first binding or interacting with a specific protein within the plant. This is known as the target site or target enzyme. Typically, inhibition of the activity of this target enzyme either directly, or more commonly indirectly, results in death of the plant (Preston and Mallory-Smith 2001). While potentially a large number of plant enzymes can interact with herbicides, in practice herbicides interact with only a small subset of these enzymes (Dayan et al. 2012; Duke 2012). This means that many herbicides can potentially inhibit the same target site, even if they have different chemical structure (Casida 2009). This has led to the organization of herbicides into a small number of distinct modes of action (Mallory-Smith and Retzinger 2003).

Herbicide resistance arises from the intensive selection pressure applied through the use of herbicides to control weedy plants (Powles and Yu 2010). Such use of herbicides removes susceptible individuals from the population leaving those that are able to survive the herbicide application. This enriches the weed population for resistance alleles and the continual use of the same herbicide or herbicides from the same mode of action results in a weed population that is not controlled by the herbicide (Jasieniuk et al. 1996).

Associate Professor (ORCID: 0000-0002-7588-124X), School of Agriculture, Food & Wine, University of Adelaide, PMB 1 Glen Osmond, South Australia 5075, Australia.
Email: christopher.preston@adelaide.edu.au

There are numerous ways that weeds can evolve resistance to herbicides; however, a common way is through a mutation that allows the target enzyme to continue to operate in the presence of the herbicide (Powles and Yu 2010). Even within target site resistance, there are several ways that a target enzyme could become resistant to herbicides. The obvious one is through a single nucleotide change within the DNA of the gene encoding the target enzyme that reduces binding of the herbicide to the target enzyme. This is typically referred to as target site resistance and has been frequently reviewed (Preston and Mallory-Smith 2001; Preston 2009; Powles and Yu 2010; Cobb and Reade 2010). More recently, other types of target site resistance have been identified. One has been gene amplification where the target gene is duplicated leading to multiple copies of the gene, more gene expression and increased target enzyme (Gaines et al. 2010). The extra copies of the target enzyme bind the herbicide, leaving sufficient uninhibited enzyme remaining for the plant to survive. There are alternative ways of achieving the same result without gene amplification. Changes in gene expression can increase the amount of target protein (Watanabe et al. 1998), as can reductions in the turnover of protein.

As most herbicides with a particular mode of action directly interact with a target protein, most commonly by binding to the protein and interfering with its function, it is plausible that one or more of the above types of target site resistance could occur in weeds under selection pressure from that mode of action (Preston 2009). The exact type of target site resistance that will be selected, or indeed whether target site resistance will be the favored mechanism, will depend on several factors including the level of resistance provided by the target site mutation compared to other mechanisms, the fitness penalty against the target site mutation and the type of selection history the population has been exposed to (Jasieniuk et al. 1996; Preston and Mallory-Smith 2001).

The only exception to this will be herbicides that do not bind to a protein, but which interact in some other way. The most obvious example is paraquat, which operates through accepting electrons from Photosystem I (PS I) (Preston 1994). This interaction is purely a redox reaction. In principle, a mutation that altered the redox potential of PS I so electrons could flow in the presence of paraquat, proving resistance. However, such a mutation would be lethal to the plant (Hawkes 2014).

Target enzyme resistance to herbicides has been reviewed extensively in recent years (Preston 2009; Powles and Yu 2010; Cobb and Reade 2010; Beckie and Tardif 2012). In this review, I will not be providing a detailed discussion of the impact of all the known mutations that have been identified in herbicide resistant weed populations for each mode of action. A list of mutations identified in field selected populations is provided in Table 1. If the reader is interested in more detail about these, they can read the several recent reviews on the topic. Recent advances in molecular biological

Table 1. Target site mutations reported for various herbicide target sites in field selected weed species.

Target Site	Amino acid residues where mutations endowing resistance occur	Amino acid substitutions reported	References
Photosystem II (PS II)	Val 219	Ile	Preston 2009
	Ser 264	Gly	
		Thr	
	Asn 266	Thr	
Acetohydroxyacid synthase (AHAS)	Ala 122	Thr	Yu and Powles 2014; Tranel et al. 2016
		Tyr	
		Val	
	Pro 197	Ala	
		Arg	
		Gln	
		His	
		Ile	
		Leu	
		Ser	
		Thr	
		Tyr	
	Ala 205	Phe	
		Val	
	Asp 376	Glu	
	Arg 377	His	
	Trp 574	Gly	
		Leu	
		Met	
	Ser 653	Asn	
		Ile	
		Thr	
	Gly 654	Asp	
		Glu	
Acetyl Coenzyme A carboxyase (ACCase)	Ile 1781	Leu	Kaundun 2014
		Val	
	Trp 1999	Cys	
		Leu	
	Trp 2027	Cys	
	Ile 2041	Asn	
		Val	
	Asp 2078	Gly	
	Cys 2088	Arg	
	Gly 2096	Ala	
		Ser	

Table 1 contd. ...

...Table 1 contd.

Target Site	Amino acid residues where mutations endowing resistance occur	Amino acid substitutions reported	References
α-tubulin	Leu 168	Phe	Preston 2009
	Thr 239	Ile	
	Met 268	Thr	
Phytoene desaturase (PDS)	Arg 304	Cys	Dayan et al. 2014
		His	
		Ser	
5-Enol-pyruvylshikimic acid 3-phosphate synthase (EPSPS)	Thr 102*	Ile	Sammons and Gaines 2014
	Pro 106	Ala	Yu et al. 2015
		Leu	
		Ser	
		Thr	
Photoporphyrinogen oxidase (PPO)	Arg 98	Leu	Dayan et al. 2014
	Gly 210**	–	
Glutamine synthetase (GS)	Asn 171	Asp	Avila-Garcia et al. 2012

* Substitutions at this site only occur in combination with Pro 106 substitutions
** Codon deletion at this site

techniques are allowing researchers greater insight into the possible changes that can occur leading to target site resistance and also into the evolutionary aspects of herbicide resistance. In this review I want to consider what we have learned in recent years about target site resistance and try to draw together the current understanding across target enzymes.

Resistance Due to Single Nucleotide Changes in Genes Coding for Target Enzymes

By far the most common target site mutations identified are single nucleotide changes within the gene encoding the target enzyme that result in a single amino acid change in the protein (Preston 2009; Powles and Yu 2010). There are several reasons why these types of mutations are so commonly reported. In the first place, they will occur readily as mutations due to simple modification of a single nucleotide (Preston 2009). Secondly, single amino acid changes that allow the enzyme to be less inhibited by the herbicide may allow the plant to survive the herbicide application and this, therefore, is a simple way for resistance to occur (Preston and Mallory-Smith 2001). However, the frequency of target site mutations may be over-stated due to the ease with which they can now be identified using modern molecular biology methods (Preston and Mallory-Smith 2001; Délye et al. 2011).

It has become evident in recent years that target site alterations as a result of single nucleotide changes, can result in more than one possible amino acid change providing resistance. There may be multiple different substitutions identified at a single amino acid within the target enzyme, multiple amino acids that can be modified within the enzyme, or both (Table 1). As most herbicides have a specific site of interaction within the target enzyme it is obvious that an amino acid modification within this site will reduce the binding of the herbicide to the target site. This means that the list of modes of action with target site mutations in weed species in Table 1 may increase in the future.

Different target site mutations will provide different levels of resistance to the herbicide (Délye et al. 2008; Preston 2009; Tranel et al. 2016). In addition, the different mutations are likely to have variable pleiotropic effects on the catalytic efficiency of the enzyme (Vila-Aiub et al. 2009; Jang et al. 2013). These pleiotropic effects may result in a measurable fitness penalty in the weeds (Jordan 1996; Tardif et al. 2006; Menchari et al. 2008). Some mutations may be rare or may not be found at all, because the pleiotropic effects on the target enzyme are so severe.

The number of possible mutations within the target enzyme can also vary. These will be dictated by the way that the herbicides interact with the target enzyme. Possible mutations are not just limited to amino acids that the herbicide directly interacts with (Preston and Mallory-Smith 2001; McCourt et al. 2006; Jang et al. 2013). Spatially adjacent amino acids can influence the environment around key binding amino acids. Changes to amino acids within the protein chain can also change the structure around the herbicide binding site and may reduce herbicide binding. For a number of target sites there are multiple mutations known that result in resistance to herbicides (Preston 2009; Yu and Powles 2010). For example, there are 7 known sites where mutations occur in acetyl-coenzyme A carboxylase (ACCase) and 8 known sites in acetohydroxyacid synthase (AHAS), also called acetolactate synthase (ALS) where mutations have been identified in weed species (Table 1). In addition, there are other possible mutation sites that have been identified through mutagenesis that have so far not appeared in weed species (Preston 2009). In contrast, all examples of target site mutations for glyphosate resistance so far reported in weed species include an amino acid modification at Pro 106 (Sammons and Gaines 2014).

The differences between target sites in the number of possible amino acid substitutions that result in resistance to herbicides are obviously related to the way the enzymes interact with both the herbicide and with the natural substrate(s) of the enzyme. X-ray crystallography of some target sites with herbicides bound to them have identified amino acids that have a close connection to the bound herbicide (Lancaster and Michel 1999; Schonbrünn et al. 2001; Pang et al. 2004; McCourt et al. 2006). Molecular modelling of target sites has also increased our understanding of the amino acids interacting with various herbicides (Anthony and Hussey 1999; Jang et al. 2013). Where the herbicide has a tight binding within the enzyme and binds to a substrate binding site, there are likely to be fewer amino acid modifications that will provide both resistance to the herbicide and a functional enzyme (Schonbrünn et al. 2001).

A good example is 5-enolpyruvylshikimate-3-phosphate synthase (EPSPS) where glyphosate acts as a transition state mimic inhibiting the binding of phosphoenol pyruvate (PEP) after the binding of shikimate-3-phosphate (Ream et al. 1992). Significant modifications of the target enzyme that completely abolish herbicide binding will likely result in an ineffective enzyme (Eschenburg et al. 2002; Funke et al. 2009). For this reason it is likely that most of the naturally-occurring mutations in EPSPS in plants are modifications of Pro 106 to Ser, Ala, Thr or Leu (Sammons and Gaines 2014). These four amino acid modifications only reduce glyphosate binding by a relatively small amount. They also reduce the efficiency of the enzyme by reducing its affinity for the substrate PEP (Healy-Fried et al. 2007). Despite the relatively low level of resistance and the effect on enzyme function, plants containing these mutations are able to survive glyphosate application and set seed.

It is clear that amino acid modifications that cause a large change within the herbicide binding pocket are likely to have a much greater impact on herbicide binding. For example, if there is a requirement for the herbicide to hydrogen bond with an amino acid within the binding pocket, then loss of that hydrogen bond might abolish the ability of the herbicide to bind at all. A good example is binding of atrazine to the plastoquinone pocket of Photosystem II (PS II). In the wild type situation, atrazine will hydrogen bond to the hydroxyl group on Ser 264 (Lancaster and Michel 1999). The mutation of Ser 264 to Gly removes this hydroxyl group and stops atrazine from binding (Ohad and Hirschberg 1990).

There may be one or several key amino acids within the herbicide binding site that the herbicide interacts with. The various interactions with the herbicide can include hydrogen bonds, ionic interactions, hydrophobic interactions and even covalent bonding (Lancaster and Michel 1999; Schonbrünn et al. 2001; Eckermann et al. 2003; McCourt et al. 2006). This means that a single mutation resulting in a single amino acid modification may just reduce herbicide binding by a small amount as the herbicide will still be interacting elsewhere within the target site.

The structure of individual target sites can also vary. Some target sites will bind the herbicides quite tightly so that only a few chemical variants can fit. A good example is EPSPS (Marzabadi et al. 1996), where there is only one commercial herbicide: glyphosate. Other target sites can accommodate chemicals of fairly widely different chemistry, such as AHAS and PS II (Tietjen et al. 1991; McCourt et al. 2006). Where many herbicides can bind to the target site, individual mutations may have widely different impacts on the binding of herbicides. While binding of some herbicides may be greatly reduced, leading to resistance, binding of others may not be affected at all or binding may even increase, leading to greater susceptibility to the herbicide (Oettmeier et al. 1991).

The various mutations within AHAS can result in varying levels of resistance to different classes of inhibitors. Mutations at Pro 197 typically result in resistance to sulfonylurea and sulfonamide herbicides, but little or no resistance to imidazolinone herbicides (Tranel et al. 2016; Preston 2009). In contrast, mutations at Ser 653 result in resistance to the imidazolinone herbicides, but not to the sulfonylurea herbicides. In addition, there can also be differences in the level of resistance between herbicides of the same chemistry. Pro 197 mutations can result in high levels of resistance to chlorsulfuron, but lower levels of resistance to triasulfuron (Preston et al. 2006).

Resistance to ACCase-inhibiting herbicides is known to be conferred by substitutions at one of seven amino acids: Ile 1781, Trp 1999, Trp 2027, Ile 2041, Asp 2078, Cys 2088, or Gly 2096 within ACCase (Kaundun 2014). Only three of these amino acids occur within 4 A of the herbicide binding (Jang et al. 2013). Residue Ile 1781 occurs close to the binding site of aryloxyphenoxypropionate, cyclohexanedione, and pinoxaden herbicides. Residues Trp 1999 and Ile 2041 occur within 4 A of the binding site for aryloxyphenoxypropionate herbicides, but not the other two chemical groups that inhibit this enzyme (Table 2). The other four amino acids are located in close proximity to amino acids within the binding pocket.

Amino acid substitutions at Ile 1785 have only been found to Val or Leu (Kaundun 2014; Jang et al. 2013). These are conservative amino acid substitutions for Ile (Jang et al. 2013). These substitutions provide low to medium levels of resistance across the three chemical classes of inhibitors of ACCase. Clearly, only small changes at 1785 are possible while retaining the activity of ACCase; however, these small changes are sufficient to reduce herbicide binding and lead to plant survival (Délye et al. 2008; Jang et al. 2013). In contrast, amino acid substitutions at Asp 2078 lead to much larger levels of resistance to all three chemical classes of inhibitor (Délye et al. 2008). Here the substitution of Asp by Gly leads to a much more significant change to the herbicide binding pocket (Jang et al. 2013).

Increasingly research into herbicide resistance in weed species is focussed on sequencing of the target enzyme looking for known mutations. Often if a previously reported mutation is not found, the assumption is made that resistance must be non-target site based. However, for several target sites, site-directed mutagenesis has identified additional mutations that have not yet been identified in weed species (Preston 2009). There may be other mutations able to provide resistance to herbicides that have not yet been identified. It is, therefore, important to conduct tests of enzyme susceptibility before non-target site resistance is declared. This problem can

Table 2. Location of amino acid residues in the binding pocket for different ACCase-inhibiting herbicides and the impact of amino acid modifications on resistance to herbicides.

Amino acid position	Amino acid present in the binding pocket*			Level of resistance to herbicides for amino acid substitutions**		
Herbicides***	APP	CHD	PXD	APP	CHD	PXD
Ile 1981	+	+	+	++	++	++
Trp 1999	+	−	−	+++	+	+
Trp 2027	−	−	−	+++	+	++
Ile 2041	+	−	−	+++	+	++
Asp 2078	−	−	−	+++	+++	+++
Cys 2088	−	−	−	++	++	++
Gly 2096	−	−	−	++	+	++

* Adapted from Jang et al. (2013). + = present in the binding site at 4 Å from the herbicide.
** Adapted from Delye et al. (2008), Kaundun et al. (2012), Cruz-Hipolito et al. (2012) and Kaundun (2014). + = resistance factor between 2 and 10, ++ = resistance factor between 11 and 100, +++ = resistance factor > 100.
*** Abbreviations: APP = aryloxyphenoxypropanoate, DIM = cyclohexanedione, PXD = pinoxaden.

be compounded by variation in the range of herbicides tested. Within most target sites there are mutations that provide resistance to only a subset of the herbicides that interact with the target site (Délye et al. 2008; Preston 2009; Tranel et al. 2016). As new molecules that inhibit the same target site are developed, these may select for amino acid modifications that hitherto have not been selected by the molecules that are in use.

Even though typically only a single amino acid modification confers resistance in most cases of target site resistance, it is certainly possible to obtain more than one mutation within a target site. The best-characterized example is with glyphosate resistance in *Eleusine indica*. Here, a Pro 106 Ser mutation is present with a Thr 102 Ile mutation in EPSPS (Yu et al. 2015). The Thr 102 Ile mutation can provide high-level resistance to the herbicide but at the expense of drastically reduced affinity for PEP (Funke et al. 2009). However, when Thr 102 Ile mutation is coupled with the Pro 106 Ser mutation, it greatly increases the level of resistance to glyphosate and at the same time maintains EPSPS efficiency (Funke et al. 2009). In this case, selection for the Thr 102 mutation likely occurred after selection for the Pro 106 mutation (Yu et al. 2015). This may have occurred through continued use of glyphosate after the first low-level resistance occurred, which then selected within that population for individuals carrying additional mutations.

A similar effect may have occurred in *Lolium rigidum* resistant to ACCase inhibiting herbicides. Malone et al. (2014) in a survey of mutations within ACCase among individuals of *L. rigidum* across a wide area of South Australia identified two individuals containing three mutations within ACCase. As *L. rigidum* is a diploid organism, two of these mutations are likely on the same allele.

Target Site Mutations in Polyploid Weed Species: A Complex Case

Many weed species are ancient polyploids containing 2 or more genomes. For these weed species target site resistance will be more complicated and harder to select with herbicides. Polyploid species will have duplicate copies of all genes. Gene silencing is a common trait observed in polyploid species, where over time the additional genes are silenced leaving a single copy active (Adams et al. 2003). This process will occur at different levels, leading to situations ranging from all genes remaining fully active, through some genes having reduced expression, to a single gene being active.

Despite these complications, target site resistance does occur in polyploid species (Christoffers et al. 2002; Yu et al. 2013; Xu et al. 2014a, 2014b). The simple case is where a mutation occurs on only a single allele, leading to a relatively low level of resistance (Yu et al. 2013; Xu et al. 2014a; Yuan et al. 2015). The more complicated example is where mutations occur on more than one of the gene copies. This will provide a higher level of resistance in the weed species (Yu et al. 2013). The higher the level of ploidy in the species, the more the difficulty of evolving target site based resistance in that species. This will be particularly true where target site mutations only provide modest levels of resistance. If all genes are equally expressed, the effect of the one target site gene containing a mutation will be diluted by the larger proportion of wild-type enzyme (Iwakami et al. 2012).

Avena fatua resistant to ACCase-inhibiting herbicides can have varying levels of resistance. Populations with target site resistance often have lower levels of resistance compared with other grass species (Christoffers et al. 2002; Cruz-Hipolito et al. 2011). The main reason is that this species has ACCase on the three genomes expressed. Mutations that are present in only one genome provide only low levels of resistance to herbicides. Individuals can have additional mutations present on other genomes and have higher levels of resistance (Yu et al. 2013). However, plants with three mutations in ACCase were still not as resistant as diploid species with the same mutations in ACCase (Yu et al. 2013).

Mutations in *EPSPS* have been identified in resistant populations of *Echinochloa colona*, another hexaploid species, with resistance to glyphosate (Han et al. 2016; Nguyen et al. 2016). In this case, the levels of resistance observed are similar to those in diploid species, such as *L. rigidum* (Wakelin and Preston 2006). At least 2 *EPSPS* genes are expressed in *E. colona* (Nguyen et al. unpublished data); so the reason why resistance is not diluted in this species by the expression of susceptible *EPSPS* is not clear.

Polyploidy also makes the identification of target site mutations through sequencing more difficult (Yu et al. 2013; Panozzo et al. 2013; Délye et al. 2015). Where the different copies of the gene are very similar, sequencing using genomic DNA may not allow the mutation to be identified, unless PCR products are cloned. This is because the genome containing the mutation may be masked by the other wild-type genomes. Equally, it might be assumed that plants are heterozygous at the resistance locus when in fact they are homozygous resistant at one locus and homozygous susceptible at the other (Xu et al. 2014b). Sequencing via cDNA will allow the identification of the expression level of the mutant alleles and provide information to estimate the contribution of the mutant allele (Panozzo et al. 2013). Next generation sequencing can also be used to identify the collection of alleles present at a specific target site (Délye et al. 2015).

Resistance due to Loss of a Codon in the Target Enzyme Gene

Theoretically, it is possible to generate target site mutations through means other than single nucleotide polymorphisms. For example, nucleotide insertions and deletions within genes can significantly change the structure of proteins. However, these typically cause frame shifts in the sequence and lead to non-functional proteins. As a result, there is so far only one example of this type of target site mutation in weed species, a codon deletion in the protoporphyrinogen oxidase (PPO) gene (Patzoldt et al. 2006).

In *Amaranthus tuberculatus*, many populations resistant to PPO inhibitors have the same Gly 201 codon deletion within the PPO gene (Thinglum et al. 2011; Schulz et al. 2015). This amino acid residue is not directly involved in binding herbicides; however, deletion of Gly 210 causes dramatic changes to the structure of the protein so that the herbicide binds less well. However, binding of the normal substrate protoporphyrinogen IX was not reduced (Hao et al. 2009; Dayan et al. 2010). Although affinity for the substrate was not affected, the mutation did greatly reduce the catalytic efficiency of the mutant PPO (Dayan et al. 2010). The Gly 210 deletion removes an important hydrogen bond to Ser 424 (Hao et al. 2009) that unravels the top of an alpha helix. This causes an expansion of the binding pocket that greatly reduces the ability of all PPO inhibitors to inhibit the activity of PPO.

It is also possible for single amino acid modifications to occur within PPO; however, only one has so far been identified in weeds. This is the substitution of Arg 98 by Leu in *Ambrosia artemisiifolia* (Rousonelos et al. 2012). The predominance of deletions at Gly 210 present in *A. tuberculatus* may be the result of the fact that amino acid modifications at other sites only provide low levels of resistance to PPO inhibitors (Horikoshi et al. 1999; Li et al. 2003; Hao et al. 2014) and this may be insufficient to provide survival under typical field applications. In addition, as each of the potential single nucleotide mutations provides resistance to only a small selection of PPO-inhibiting herbicides (Hao et al. 2014), the PPO-inhibiting herbicides used may not select for these single amino acid modifications.

Resistance due to Gene Amplification of the Target Enzyme Gene

A completely different way that target site-based resistance can occur is through increasing the number of gene copies. Gene amplification as a resistance mechanism had been identified in cell culture with several herbicides (Donn et al. 1984; Smart et al. 1985; Deak et al. 1988; Goldsbrough et al. 1990; Harms et al. 1992), but was only first described in weeds with glyphosate resistance in *Amaranthus palmeri* (Gaines et al. 2010). Individuals of this species that were resistant to glyphosate had up to 160 times as many copies of the target enzyme EPSPS compared with susceptible plants. There was a corresponding increase in the amount of EPSPS message produced and in EPSPS activity (Gaines et al. 2010). The plants survived glyphosate application because the extra protein was able to bind the glyphosate leaving some enzyme unaffected by the herbicide.

Since then gene amplification of EPSPS has been identified in several other species including *A. tuberculatus* (Tranel et al. 2010; Chatham et al. 2015), *A. spinosis* (Nandula et al. 2014), *Kochia scoparia* (Wiersma et al. 2015) and two grass species: *Lolium perenne* ssp. *multiflorum* (Salas et al. 2012) and *Bromus diandrus* (Malone et al. 2016). Different approaches to gene amplification have been identified in different species. *A. palmeri*, and *B. diandrus* seem to have many copies of *EPSPS* scattered around the genome (Gaines et al. 2010; Malone et al. 2016). *K. scoparia* on the other hand seems to have a single set of tandem repeats (Jugulam et al. 2014). These differences may simply be due to chance; however, they may result from the number of extra copies required to achieve resistance and the limits to tandem repeats. In *K. scoparia* > 3 copies of the gene are sufficient to allow the plant to survive a field application of glyphosate (Wiersma et al. 2014). In the other species, many more copies may be required and so a different pattern of gene amplification occurs (Gaines et al. 2011; Malone et al. 2016). However, lower copy number variants of *A. palmeri* resistant to glyphosate have also been reported (Mohseni-Moghadam et al. 2013).

Where it has been examined in detail, gene amplification has occurred with only a single EPSPS allele. In *A. palmeri*, only one of the two EPSPS alleles is amplified (Gaines et al. 2013). Likewise, in *K. scoparia*, sequence evidence indicates the presence of a second EPSPS allele in resistant plants that is not amplified (Wiersma et al. 2015). In *B. diandrus*, at least three EPSPS alleles were identified in glyphosate susceptible plants with one of these preferentially amplified in resistant plants from two different populations (Malone et al. 2016).

Gene amplification has not yet been demonstrated for any other herbicide mode of action. It may be demonstrated in the future. This raises some interesting questions as to why this mechanism is so prevalent with glyphosate resistance only. The answer may lie in the relative weakness of target site mutations in *EPSPS* (Sammons and Gaines 2014), making them harder to select in specific weed species. Therefore, the rarer gene amplification mechanism is favoured.

Conclusions

Target site resistance to herbicides is an easy way that weed species can evolve to survive herbicide application. While most examples of target site resistance are simple single nucleotide changes in the gene, leading to an amino acid modification in the protein, it has become obvious in recent years that other types of target site mutations are also possible. These include codon deletions in PPO and gene amplification of *EPSPS*. Many, but not all, target site mutations will have detrimental effects on the functioning of the target enzyme. Therefore, individuals carrying these mutations tend to be rare in unselected weed populations.

For many target sites, more than one amino acid modification providing resistance to herbicides is known. The frequency with which specific amino acid modifications are observed will be a complex interplay between the fitness impacts of each mutation, the level of resistance provided to individual selecting herbicides, the innate tolerance of the weed species to the herbicide and chance. This means that different mutations may be predominantly selected in different locations based entirely on herbicide preferences (Menchari et al. 2006; Malone et al. 2014), and may change over time (Rosenhauer et al. 2013; Malone et al. 2014). Alternatively, different mutations may be favoured in different weed species treated with the same herbicide (Malone et al. 2014).

The situation can be complicated by polyploidy in weed species. In this case, resistance is influenced by gene dosage effects, dilution of the resistance allele by susceptible alleles, the relative expression of the various alleles and on which of the alleles the mutation occurs. A mutation that provides high levels of resistance in a diploid species may provide much lower levels of resistance in a polyploid species (Yu et al. 2013). Due to the dilution effects of susceptible alleles, selection in weed species with higher ploidy levels may not favour target site resistance. Allelic dilution of target site resistance is not observed in all examples of polyploid weed species, suggesting there remains much to learn about target site resistance in polyploid weed species.

The widespread availability of inexpensive molecular biology tools has played a crucial role in identifying target site resistance in weed species. However, as these tools are so readily available, they risk a situation where this is the only work done to investigate herbicide resistance in weeds. It is becoming increasingly obvious that several common weed species have more than one mechanism of resistance to many herbicides, and are more difficult to identify so some mechanisms may be overlooked (Preston and Mallory-Smith 2001; Délye et al. 2011). In addition, simply identifying a mutation within a target site does little to aid the understanding of selection for resistance. Research to understand the specific impacts of individual mutations on the structure and activity of the target enzyme is crucial in putting together an understanding of the complex factors involved in selection.

LITERATURE CITED

Adams KL, Cronn R, Percifield R and Wendel JF (2003) Genes duplicated by polyploidy show unequal contributions to the transcriptome and organ-specific reciprocal silencing. Proc. Natl. Acad. Sci. USA 100: 4649–4654.

Anthony RG and Hussey PJ (1999) Double mutation in *Eleusine indica* α-tubulin increases the resistance of transgenic maize calli to dinitroaniline and phosphorothioamidate herbicides. Plant J. 18: 669–674.

Avila-Garcia WV, Sanchez-Olguin E, Hulting AG and Mallory-Smith C (2012) Target-site mutation associated with glufosinate resistance in Italian ryegrass (*Lolium perenne* L. ssp. *multiflorum*). Pest Man. Sci. 68: 1248–1254.

Beckie HJ and Tardif FJ (2012) Herbicide cross resistance in weeds. Crop Prot. 35: 15–28.

Casida JE (2009) Pest toxicology: the primary mechanisms of pesticide action. Chem. Res. Toxicol. 22: 609–619.

Chatham LA, Wu C, Riggins CW, Hager AG, Young BG, Roskamp GK and Tranel PJ (2015) EPSPS gene amplification is present in the majority of glyphosate-resistant Illinois waterhemp (*Amaranthus tuberculatus*) populations. Weed Technol. 29: 48–55.

Christoffers MJ, Berg ML and Messersmith CG (2002) An isoleucine to leucine mutation in acetyl-CoA carboxylase confers herbicide resistance in wild oat. Genome 45: 1049–1056.

Cobb AH and Reade JPH (2010) Herbicide resistance. pp. 216–237. *In*: Cobb AH and Reade JPH (eds.). Herbicides and Plant Physiology, Second Edition, Oxford, UK: Wiley-Blackwell.

Cruz-Hipolito H, Osuna MD, Dominguez-Valenzuela JA, Espinoza N and De Prado R (2011) Mechanism of resistance to ACCase-inhibiting herbicides in wild oat (*Avena fatua*) from Latin America. J. Agric. Food Chem. 59: 7261–7267.

Cruz-Hipolito H, Domínguez-Valenzuela JA, Osuna MD and De Prado R (2012) Resistance mechanism to acetyl coenzyme A carboxylase inhibiting herbicides in *Phalaris paradoxa* collected in Mexican wheat fields. Plant Soil 355: 121–130.

Dayan FE, Daga PR, Duke SO, Lee RM, Tranel PJ and Doerksen RJ (2010) Biochemical and structural consequences of a glycine deletion in the α-8 helix of protoporphyrinogen oxidase. Biochim. Biophys Acta 1804: 1548–1556.

Dayan FE, Owens DK and Duke SO (2012) Rationale for a natural products approach to herbicide discovery. Pest Manag. Sci. 68: 519–528.

Dayan FE, Owens DK, Tranel PJ, Preston C and Duke SO (2014) Evolution of resistance to phytoene desaturase and protoporphyrinogen oxidase inhibitors—state of knowledge. Pest Manag. Sci. 70: 1358–1366.

Deak M, Donn G, Feher A and Dudits D (1988) Dominant expression of a gene amplification related herbicide resistance in *Medicago* cell hybrids. Plant Cell Rep. 7: 158–161.

Délye C, Matéjicek A and Michel S (2008) Cross-resistance patterns to ACCase-inhibiting herbicides conferred by mutant ACCase isoforms in *Alopecurus myosuroides* Huds. (black-grass), re-examined at the recommended herbicide field rate. Pest Manag. Sci. 64: 1179–1186.

Délye C, Gardin JAC, Boucansaud K, Chauvel B and Petit C (2011) Non-target-site-based resistance should be the centre of attention for herbicide resistance research: *Alopecurus myosuroides* as an illustration. Weed Res. 51: 433–437.

Délye C, Causse R, Gautier V, Poncet C and Michel S (2015) Using next-generation sequencing to detect mutations endowing resistance to pesticides: application to acetolactate-synthase (ALS)-based resistance in barnyard grass, a polyploid grass weed. Pest Manag. Sci. 71: 675–685.

Donn G, Tischer E, Smith JA and Goodman HM (1984) Herbicide-resistant alfalfa cells: an example of gene amplification in plants. J. Mol. Appl. Genet. 2: 621–635.

Duke SO (2012) Why have no new herbicide modes of action appeared in recent years? Pest Manag. Sci. 68: 505–512.

Eckermann C, Matthes B, Nimtz M, Reiser V, Lederer B, Böger P and Schröder J (2003) Covalent binding of chloroacetamide herbicides to the active site cysteine of plant type III polyketide synthases. Phytochemistry 64: 1045–1054.

Eschenburg S, Healy ML, Priestman MA, Lushington GH and Schönbrunn E (2002) How the mutation glycine96 to alanine confers glyphosate insensitivity to 5-enolpyruvyl shikimate-3-phosphate synthase from *Escherichia coli*. Planta 216: 129–135.

Funke T, Yang Y, Han H, Healy-Fried M, Olesen S, Becker A and Schönbrunn E (2009) Structural basis of glyphosate resistance resulting from the double mutation Thr97→ Ile and Pro101→ Ser in 5-enolpyruvylshikimate-3-phosphate synthase from *Escherichia coli*. J. Biol. Chem. 284: 9854–9860.

Gaines TA, Shaner DL, Ward SM, Leach JE, Preston C and Westra P (2011) Mechanism of resistance of evolved glyphosate-resistant Palmer amaranth (*Amaranthus palmeri*). J. Agric. Food Chem. 59: 5886–5889.

Gaines TA, Wright AA, Molin WT, Lorentz L, Riggins CW, Tranel PJ, Beffa R, Westra P and Powles SB (2013) Identification of genetic elements associated with EPSPS gene amplification. PloS One 8: 10.1371/journal.pone.0065819.

Gaines TA, Zhang W, Wang D, Bukun B, Chisholm ST, Shaner DL, Nissen SJ, Patzoldt WL, Tranel PJ, Culpepper AS, Grey TL, Webster TM, Vencill WK, Sammons RD, Jiang J, Preston C, Leach JE and Westra P (2010) Gene amplification confers glyphosate resistance in *Amaranthus palmeri*. Proc. Natl. Acad. Sci. USA 107: 1029–1034.

Goldsbrough PB, Hatch EM, Huang B, Kosinski WG, Dyer WE, Herrmann KM and Weller SC (1990) Gene amplification in glyphosate tolerant tobacco cells. Plant Sci. 72: 53–62.

Han H, Yu Q, Widderick MJ and Powles SB (2016) Target-site EPSPS Pro-106 mutations: sufficient to endow glyphosate resistance in polyploid *Echinochloa colona*? Pest Manag. Sci. 72: 264–271.

Hao GF, Zhu XL, Ji FQ, Zhang L, Yang GF and Zhan CG (2009) Understanding the mechanism of drug resistance due to a codon deletion in protoporphyrinogen oxidase through computational modeling. J. Phys. Chem. B 113: 4865–4875.

Hao GF, Tan Y, Xu WF, Cao RJ, Xi Z and Yang GF (2014) Understanding resistance mechanism of protoporphyrinogen oxidase-inhibiting herbicides: Insights from computational mutation scanning and site-directed mutagenesis. J. Agric. Food Chem. 62: 7209–7215.

Harms CT, Armour SL, DiMaio JJ, Middlesteadt LA, Murray D, Negrotto DV, Thompson-Taylor H, Weymann K, Montoya AL, Shillito RD and Jen GC (1992) Herbicide resistance due to amplification of a mutant acetohydroxyacid synthase gene. Mol. Gen. Genet. 233: 427–435.

Hawkes TR (2014) Mechanisms of resistance to paraquat in plants. Pest Manag. Sci. 70: 1316–1323.

Healy-Fried ML, Funke T, Priestman MA, Han H and Schönbrunn E (2007) Structural basis of glyphosate tolerance resulting from mutations of Pro101 in *Escherichia coli* 5-enolpyruvylshikimate-3-phosphate synthase. J. Biol. Chem. 282: 32949–32955.

Horikoshi M, Mametsuka K and Hirooka T (1999) Molecular basis of photobleaching herbicide resistance in tobacco. J. Pestic. Sci. 24: 17–22.

Iwakami S, Uchino A, Watanabe H, Yamasue Y and Inamura T (2012) Isolation and expression of genes for acetolactate synthase and acetyl-CoA carboxylase in *Echinochloa phyllopogon*, a polyploid weed species. Pest Manag. Sci. 68: 1098–1106.

Jang S, Marjanovic J and Gornicki P (2013) Resistance to herbicides caused by single amino acid mutations in acetyl-CoA carboxylase in resistant populations of grassy weeds. New Phytol. 197: 1110–1116.

Jasieniuk M, Brûlé-Babel AL and Morrison IN (1996) The evolution and genetics of herbicide resistance in weeds. Weed Sci. 44: 176–193.

Jordan N (1996) Effects of the triazine-resistance mutation on fitness in *Amaranthus hybridus* (smooth pigweed). J. Appl. Ecol. 33: 141–150.

Jugulam M, Niehues K, Godar AS, Koo DH, Danilova T, Friebe B, Sehgal S, Varanasi VK, Wiersma A, Westra P and Stahlman PW (2014) Tandem amplification of a chromosomal segment harboring 5-enolpyruvylshikimate-3-phosphate synthase locus confers glyphosate resistance in *Kochia scoparia*. Plant Physiol. 166: 1200–1207.

Kaundun SS (2014) Resistance to acetyl-CoA carboxylase-inhibiting herbicides. Pest Manag. Sci. 70: 1405–1417.

Kaundun SS, Hutchings SJ, Dale RP and McIndoe E (2012) Broad resistance to ACCase inhibiting herbicides in a ryegrass population is due only to a cysteine to arginine mutation in the target enzyme. PloS One 7: 10.1371/journal.pone.0039759.

Lancaster CRD and Michel H (1999) Refined crystal structures of reaction centres from *Rhodopseudomonas viridis* in complexes with the herbicide atrazine and two chiral atrazine derivatives also lead to a new model of the bound carotenoid. J. Mol. Biol. 286: 883–898.

Li X, Volrath SL, Nicholl DB, Chilcott CE, Johnson MA, Ward ER and Law MD (2003) Development of protoporphyrinogen oxidase as an efficient selection marker for *Agrobacterium tumefaciens*-mediated transformation of maize. Plant Physiol. 133: 736–747.

McCourt JA, Pang SS, King-Scott J, Guddat LW and Duggleby RG (2006) Herbicide-binding sites revealed in the structure of plant acetohydroxyacid synthase. Proc. Natl. Acad. Sci. USA 103: 569–573.

Mallory-Smith CA and Retzinger Jr EJ (2003) Revised classification of herbicides by site of action for weed resistance management strategies. Weed Technol. 17: 605–619.

Malone JM, Boutsalis P, Baker J and Preston C (2014) Distribution of herbicide-resistant acetyl-coenzyme A carboxylase alleles in *Lolium rigidum* across grain cropping areas of South Australia. Weed Res. 54: 78–86.

Malone JM, Morran S, Shirley N, Boutsalis P and Preston C (2016) EPSPS gene amplification in glyphosate-resistant *Bromus diandrus*. Pest Man. Sci. 72: 81–88.

Marzabadi MR, Gruys KJ, Pansegrau PD, Walker MC, Yuen HK and Sikorski JA (1996) An EPSP synthase inhibitor joining shikimate 3-phosphate with glyphosate: synthesis and ligand binding studies. Biochemistry 35: 4199–4210.

Menchari Y, Camilleri C, Michel S, Brunel D, Dessaint F, Le Corre V and Délye C (2006) Weed response to herbicides: regional-scale distribution of herbicide resistance alleles in the grass weed *Alopecurus myosuroides*. New Phytol. 171: 861–874.

Menchari Y, Chauvel B, Darmency H and Délye C (2008) Fitness costs associated with three mutant acetyl-coenzyme A carboxylase alleles endowing herbicide resistance in blackgrass *Alopecurus myosuroides*. J. Appl. Ecol. 45: 939–947.

Mohseni-Moghadam M, Schroeder J and Ashigh J (2013) Mechanism of resistance and inheritance in glyphosate resistant Palmer amaranth (*Amaranthus palmeri*) populations from New Mexico, USA. Weed Sci. 61: 517–525.

Nandula VK, Wright AA, Bond JA, Ray JD, Eubank TW and Molin WT (2014) EPSPS amplification in glyphosate-resistant spiny amaranth (*Amaranthus spinosus*): a case of gene transfer via interspecific hybridization from glyphosate-resistant Palmer amaranth (*Amaranthus palmeri*). Pest Manag. Sci. 70: 1902–1909.

Nguyen TH, Malone JM, Boutsalis P, Shirley N and Preston C (2016) Temperature influences the level of glyphosate resistance in barnyardgrass (*Echinochloa colona*). Pest Manag. Sci. 72: 1031–1039.

Ohad N and Hirschberg J (1990) A similar structure of the herbicide binding site in photosystem II of plants and cyanobacteria is demonstrated by site specific mutagenesis of the psbA gene. Photosynth. Res. 23: 73–79.

Oettmeier W, Hilp U, Draber W, Fedtke C and Schmidt RR (1991) Structure-activity relationships of triazinone herbicides on resistant weeds and resistant *Chlamydomonas reinhardtii*. Pestic. Sci. 33: 399–409.

Pang SS, Guddat LW and Duggleby RG (2004) Crystallization of *Arabidopsis thaliana* acetohydroxyacid synthase in complex with the sulfonylurea herbicide chlorimuron ethyl. Acta Cryst. D 60: 153–155.

Panozzo S, Scarabel L, Tranel PJ and Sattin M (2013) Target-site resistance to ALS inhibitors in the polyploid species *Echinochloa crus-galli*. Pestic. Biochem. Physiol. 105: 93–101.

Patzoldt WL, Hager AG, McCormick JS and Tranel PJ (2006) A codon deletion confers resistance to herbicides inhibiting protoporphyrinogen oxidase. Proc. Natl. Acad. Sci. USA 103: 12329–12334.

Powles SB and Yu Q (2010) Evolution in action: plants resistant to herbicides. Annu. Rev. Plant Biol. 61: 317–347.

Preston C (1994) Resistance to photosystem I inhibiting herbicides. pp. 61–82. *In*: Powles SB and Holtum JAM (eds.). Herbicide Resistance in Plants: Biology and Biochemistry. Boca Raton, FL: Lewis Publishers.

Preston C (2009) Herbicide resistance: target site mutations. pp. 127–148. *In*: Stewart Jr CN (ed.). Weedy and Invasive Plant Genomics. Ames IA: Wiley-Blackwell.

Preston C and Mallory-Smith CA (2001) Biochemical mechanisms, inheritance, and molecular genetics of herbicide resistance in weeds. pp. 23–60. *In*: Powles SB and Shaner DL (eds.). Herbicide Resistance and World Grains. Boca Raton, FL: CRC Press.

Preston C, Stone LM, Rieger MA and Baker J (2006) Multiple effects of a naturally occurring proline to threonine substitution within acetolactate synthase in two herbicide-resistant populations of *Lactuca serriola*. Pestic. Biochem. Physiol. 84: 227–235.

Ream JE, Yuen HK, Frazier RB and Sikorski JA (1992) EPSP synthase: binding studies using isothermal titration microcalorimetry and equilibrium dialysis and their implications for ligand recognition and kinetic mechanism. Biochemistry 31: 5528–5534.

Rosenhauer M, Jaser B, Felsenstein FG and Petersen J (2013) Development of target-site resistance (TSR) in *Alopecurus myosuroides* in Germany between 2004 and 2012. J. Plant Dis. Prot. 120: 179–187.

Rousonelos SL, Lee RM, Moreira MS, VanGessel MJ and Tranel PJ (2012) Characterization of a common ragweed (*Ambrosia artemisiifolia*) population resistant to ALS and PPO-inhibiting herbicides. Weed Sci. 60: 335–344.

Salas RA, Dayan FE, Pan Z, Watson SB, Dickson JW, Scott RC and Burgos NR (2012) EPSPS gene amplification in glyphosate-resistant Italian ryegrass (*Lolium perenne* ssp. *multiflorum*) from Arkansas. Pest Manag. Sci. 68: 1223–1230.

Sammons RD and Gaines TA (2014) Glyphosate resistance: state of knowledge. Pest Manag. Sci. 70: 1367–1377.

Schönbrunn E, Eschenburg S, Shuttleworth WA, Schloss JV, Amrhein N, Evans JN and Kabsch W (2001) Interaction of the herbicide glyphosate with its target enzyme 5-enolpyruvylshikimate 3-phosphate synthase in atomic detail. Proc. Natl. Acad. Sci. USA 98: 1376–1380.

Schultz JL, Chatham LA, Riggins CW, Tranel PJ and Bradley KW (2015) Distribution of herbicide resistances and molecular mechanisms conferring resistance in Missouri waterhemp (*Amaranthus rudis* Sauer) populations. Weed Sci. 63: 336–345.

Smart CC, Johänning D, Müller G and Amrhein N (1985) Selective overproduction of 5-enol-pyruvylshikimic acid 3-phosphate synthase in a plant cell culture which tolerates high doses of the herbicide glyphosate. J. Biol. Chem. 260: 16338–16346.

Tardif FJ, Rajcan I and Costea M (2006) A mutation in the herbicide target site acetohydroxyacid synthase produces morphological and structural alterations and reduces fitness in *Amaranthus powellii*. New Phytol. 169: 251–264.

Thinglum KA, Riggins CW, Davis AS, Bradley KW, Al-Khatib K and Tranel PJ (2011) Wide distribution of the waterhemp (*Amaranthus tuberculatus*) ΔG210 PPX2 mutation, which confers resistance to PPO-inhibiting herbicides. Weed Sci. 59: 22–27.

Tietjen KG, Kluth JF, Andree R, Haug M, Lindig M, Müller KH, Wroblowsky HJ and Trebst A (1991) The herbicide binding niche of photosystem II—a model. Pestic. Sci. 31: 65–72.

Tranel PJ, Riggins CW, Bell MS and Hager AG (2010) Herbicide resistances in *Amaranthus tuberculatus*: a call for new options. J. Agric. Food Chem. 59: 5808–5812.

Tranel PJ, Wright TR and Heap IM (2016) Mutations in herbicide-resistant weeds to ALS inhibitors. Online http://www.weedscience.org/mutations/mutationdisplayall.aspx Accessed: 20 January, 2016.

Vila-Aiub MM, Neve P and Powles SB (2009) Fitness costs associated with evolved herbicide resistance alleles in plants. New Phytol. 184: 751–767.

Wakelin AM and Preston C (2006) A target-site mutation is present in a glyphosate-resistant *Lolium rigidum* population. Weed Res. 46: 432–440.

Watanabe N, Che FS, Iwano M, Takayama S, Nakano T, Yoshida S and Isogai A (1998) Molecular characterization of photomixotrophic tobacco cells resistant to protoporphyrinogen oxidase-inhibiting herbicides. Plant Physiol. 118: 751–758.

Wiersma AT, Gaines TA, Preston C, Hamilton JP, Giacomini D, Buell CR, Leach JE and Westra P (2015) Gene amplification of 5-enol-pyruvylshikimate-3-phosphate synthase in glyphosate-resistant *Kochia scoparia*. Planta 241: 463–474.

Xu H, Li J, Zhang D, Cheng Y, Jiang Y and Dong L (2014a) Mutations at codon position 1999 of acetyl-CoA carboxylase confer resistance to ACCase-inhibiting herbicides in Japanese foxtail (*Alopecurus japonicus*). Pest Man. Sci. 70: 1894–1901.

Xu H, Zhang W, Zhang T, Li J, Wu X and Dong L (2014b) Determination of ploidy level and isolation of genes encoding acetyl-CoA carboxylase in Japanese foxtail (*Alopecurus japonicus*). PloS One 9: 10.1371/journal.pone.0114712.

Yu Q and Powles SB (2014) Resistance to AHAS inhibitor herbicides: current understanding. Pest Man. Sci. 70: 1340–1350.

Yu Q, Ahmad-Hamdani MS, Han H, Christoffers MJ and Powles SB (2013) Herbicide resistance-endowing ACCase gene mutations in hexaploid wild oat (*Avena fatua*): insights into resistance evolution in a hexaploid species. Heredity 110: 220–231.

Yu Q, Jalaludin A, Han H, Chen M, Sammons RD and Powles SB (2015) Evolution of a double amino acid substitution in the 5-enolpyruvylshikimate-3-phosphate synthase in *Eleusine indica* conferring high-level glyphosate resistance. Plant Physiol. 167: 1440–1447.

Yuan G, Liu W, Bi Y, Du L, Guo W and Wang J (2015) Molecular basis for resistance to ACCase-inhibiting herbicides in *Pseudosclerochloa kengiana* populations. Pestic. Biochem. Physiol. 119: 9–15.

Chapter 10

Gene Amplification and Herbicide Resistance

Application of Molecular Cytogenetic Tools

Mithila Jugulam,[1,*] *Karthik Putta,*[1] *Vijay K. Varanasi*[1] *and Dal-Hoe Koo*[2]

Introduction

Herbicides have made significant contributions to modern agriculture by offering exceptional weed management in crops and also facilitate no-till crop production to conserve soil and moisture. However, repeated field application of herbicides with the same mode of action has resulted in the selection of herbicide-resistant weeds. Mechanisms which confer resistance to herbicides can broadly be categorized into two types: (a) non-target site resistance (NTSR) and (b) target site resistance (TSR) (discussed in greater detail elsewhere in this book). Briefly, NTSR mechanisms include reduced herbicide uptake/translocation, and/or enhanced herbicide detoxification, decreased rates of herbicide activation, or sequestration of the herbicide (Devine and Eberlein 1997). On the other hand, TSR, essentially involves any alteration in the herbicide target site, such as mutations in target gene affecting herbicide binding kinetics (Powles and Yu 2010) or as more recently reported in glyphosate-resistant weeds, amplification of target gene (Sammons and Gaines 2014).

Gene amplification primarily implies the replication of a copy of a gene (Bass and Field 2011). Gene amplification is a major impetus for the

[1] Department of Agronomy, Kansas State University, Manhattan, KS.
[2] Department of Plant Pathology, Kansas State University, Manhattan, KS.
* Corresponding author: mithila@ksu.edu

creation of genetic diversity in nature (Wagner et al. 2007). In weed species, till date, gene amplification as a mechanism of herbicide resistance has been documented only for glyphosate. Glyphosate non-selectively inhibits 5-enolpyruvylshikimate-3-phosphate synthase (EPSPS), preventing the biosynthesis of the aromatic amino acids phenylalanine, tyrosine, and tryptophan (Steinrücken and Amrbein 1980) resulting in plant death. As a result of extensive and exclusive use, about 35 weed species across the globe have evolved resistance to glyphosate (Heap 2016). In addition to *EPSPS* gene amplification, glyphosate resistance has also been reported to have evolved as a result of reduced translocation and/or sequestration of glyphosate (Robertson 2010; Segobye 2013; Preston and Wakelin 2008) or mutation in the *EPSPS* gene resulting in substitution of proline 106 to alanine, leucine, serine, or threonine, causing a change in the binding site of the EPSPS enzyme (Powles and Preston 2006; Sammons and Gaines 2014; Shaner et al. 2012). Recently, in goosegrass (*Eleusine indica*) a double mutation in the *EPSPS* gene with substitution of proline 106 to serine and threonine 102 to isoleucine, T102I + P106S (TIPS) was found to confer 180 fold more resistance to glyphosate when compared to susceptible plants (Yu et al. 2015). Nonetheless, amplification of the *EPSPS* gene appears to be the basis for glyphosate resistance in several weeds (Sammons and Gaines 2014).

Over the last three decades, tremendous advancements have been made in the molecular biology and genomics of plants including weed species. Whole genome sequence information of a number of plant species has been made available. Molecular cytogenetics is a discipline which encompasses the use of molecular tools to study various aspects of chromosome behavior and karyotyping. The genome sequence data allows comparison of DNA from different species through coding regions, introns, and repetitive gene sequences. On the other hand, progress made in cytogenetics, especially by using molecular tools helped immensely to identify the physical location of genes present on the chromosomes. Application of molecular cytogenetics tools will facilitate bridging the gap between our understanding of the genome at molecular level and that of at cytogenetics level. Molecular cytogenetics enables precise analysis of chromosomes to understand genetics, genetic recombination, and karyotyping of the genes. Significant advances have been made in the development of cytogenetic maps using molecular tools to identify the location of genes on the chromosomes in plant species (Cui et al. 2015). Direct *in situ* hybridization of labeled DNA probes of the gene(s) of interest on the chromosomes is possible using molecular cytogenetic tools. Physical mapping of genes of interest has been accomplished in a number of economically important crop species (Amarillo and Bass 2007; Howell et al. 2005; Xiong et al. 2010). However, only recently the molecular cytogenetic tools were employed in weed science, specifically, to locate the amplified copies of *EPSPS* gene in glyphosate-resistant weed species (Gaines et al. 2010; Jugulam et al. 2014). The aim of this chapter is to provide an overview of gene amplification as a mechanism of herbicide resistance and to discuss the scope of use of molecular cytogenetic tools in weed science.

Gene Amplification and Resistance to Xenobiotics

Gene amplification conferring resistance to xenobiotics has been documented extensively in arthropods and eukaryotic cancer cells (Powles 2010). It has been suggested that the amplification of genes can be induced when organisms are subjected to biotic or abiotic stresses (Slack et al. 2006). Bass and Field (2011) suggested that the pesticide resistance in arthropods typically develops via amplification of genes coding for esterases, glutathione S-transferases, or cytochrome P450 monooxygenases, which are known to detoxify or inactivate a variety of pesticides. Peach-potato aphid (*Myzus persicae*) was found to be resistant to many insecticides due to amplification of the gene encoding for esterase E4 that can metabolize the insecticides (Field et al. 1988; Field and Devonshire 1997). Similarly, *Culex* mosquitoes resistant to organophosphorus compounds were found to have up to 250 copies of the genes coding for esterase B1 and B2 (Mouches et al. 1986; Paton et al. 2000; Raymond et al. 1993). Several cases of plant resistance to fungicides and insecticides have also been attributed to amplification of genes involved in pesticide detoxification (Devonshire and Field 1991; Ma and Michailides 2005). However, the first case of gene amplification in response to herbicide selection *in vivo* was reported in alfalfa (*Medicago sativa*) (Donn et al. 1984). Alfalfa cell lines with 4–11 fold amplification of the gene coding for *glutamine synthetase* (*GS*), the target site of herbicide glufosinate were found in response to selection. These lines were 20–100 fold more resistant to glufosinate compared to the wild-type (Donn et al. 1984).

The first case of naturally evolved resistance to glyphosate, via gene amplification was documented in Palmer amaranth (*Amaranthus palmeri*). Glyphosate-resistant Palmer amaranth had 5 to 160 fold more copies of *EPSPS* gene compared to susceptible individuals (Gaines et al. 2010). Subsequently, *EPSPS* gene amplification resulting in glyphosate resistance has been reported in other weed species. For example, glyphosate-resistant kochia (*Kochia scoparia*) and waterhemp (*Amaranthus tuberculatus*) plants had up to 3–15 more *EPSPS* copies than the susceptible pants (Jugulam et al. 2014; Wiersma et al. 2015; Chatham et al. 2015). *EPSPS* gene amplification as a mechanism of glyphosate resistance has also been reported in the monocot weeds, such as Italian ryegrass (*Lolium multiflorum*) (Salas et al. 2012), goosegrass (Chen et al. 2015), and ripgut brome (*Bromus diandrus*) (Malone et al. 2016). In all these species, glyphosate-resistant plants had > 20 *EPSPS* copies compared to the susceptible individuals (Salas et al. 2012; Chen et al. 2015; Malone et al. 2016).

Basis of Gene Amplification

Gene amplification essentially results in a selective increase in the number of copies of a gene, more likely without the proportional increase of other genes in a given genome. Ohno (1970) first hypothesized that organisms can evolve gene duplication as an impetus to produce more of the same product or generate new genetic loci with novel gene functions. The term gene amplification, gene duplication, or chromosomal duplication can all be used synonymously. Gene amplification is a precursor for genetic diversity and occurs naturally in several organisms (Wagner et al. 2007). In plant species, amplification of pre-existing genes contributes significantly to the genetic diversity (Flagel and Wendel 2009). Some duplicated genes undergo further amplification under selection, providing an immediate adaptive advantage to withstand stress (Perry et al. 2007). It appears that without gene amplification, the plasticity of any species in adapting to changing biotic and abiotic (i.e., herbicides) stress factors would be severely impaired. Furthermore, in plants, gene amplification may have facilitated the evolution of complex gene expression networks. The importance of gene duplication or amplification as a prominent event in driving biological evolution has been reported in many instances (Soukup 1974; Taylor and Raes 2004; Kubo et al. 2015; Zhang 2003).

Gene duplications can arise through several mechanisms, such as, (1) polyploidy, (2) trisomy, (3) unequal crossing over, and (4) mediated via transposons. Polyploidy (genome duplication) refers to an increase in the genome copies resulting in a corresponding increase in the gene copy number and possible transcriptional up-regulation of the genes (Schoenfelder and Fox 2015). Trisomy, on the other hand, occurs as a result of the formation of an extra chromosome in aneuploid cells. The significance of the other two mechanisms, i.e., unequal cross over or transposon-mediated gene amplification has been implicated in the evolution of glyphosate resistance in weed species. These two mechanisms are discussed in more detail below.

Crossing over is a naturally occurring process in which DNA segments are exchanged between two perfectly arranged homologous chromosomes (paternal and maternal) during meiosis I. However, occasionally, the presence of repetitive sequences can misalign the two homologous chromosomes, leading to unequal crossover event resulting in the variation in the gene copy number (Eichler 2008). Unequal crossing over or nonreciprocal recombination results in duplication of one chromosome segment and deletion of another. The initial event of the *EPSPS* gene duplication in glyphosate-resistant kochia may have occurred as a result of unequal recombination, because, we (Jugulam et al. 2014) demonstrated tandem arrangement of *EPSPS* copies in glyphosate-resistant kochia. In reponse to glyphosate selection, continuous variation in *EPSPS* copy number, and a positive correlation between *EPSPS* expression and the copy number was also reported in kochia (Jugulam et al. 2014; Wiersma et al. 2015) suggesting that the *EPSPS* copy number in kochia increases through an adaptive process. Furthermore, we also illustrated the hybridization of *EPSPS* probes at distal ends of homologous chromosomes of kochia (Jugulam et al. 2014) implying that increase in *EPSPS* copies in glyphosate-resistant kochia may have occurred as a result of unequal cross over, primarily because, the gene duplication via unequal cross over likely occurs at the telomeric region of the chromosomes (Royle et al. 1988; Amarger et al. 1998; Ames et al. 2008). It is still unknown if the initial event of duplication of *EPSPS* copies in kochia has occurred before glyphosate selection was imposed. The significance of unequal recombination in the formation of many disease resistance gene clusters in crop plants has been documented (Van der Hoorn et al. 2001; Nagy and Bennetzen 2008; Luo et al. 2011).

Transposable elements (or transposons) are DNA sequences creating genetic variation through random movement from one location to another in the genome (Federoff 2012). Transposons are activated under stress conditions leading to gene amplification (Lisch 2009). Transposons were first discovered in maize by Barbara McClintock (1951), and it was suggested that these genetic elements are constantly subjected to alteration and rearrangement in the genome. Initially thought as "junk" DNA with no specific function, transposable elements are now known to have several regulatory functions (Muotri et al. 2007). There are two major classes of transposable elements commonly found in eukaryotic organisms: class I (RNA elements) and class II (DNA elements) (Kejnovsky et al. 2012). Both these classes of transposable elements may result in an increase in genomic copy number and gene amplification. Class I transposable elements (retrotransposons) such as long interspersed nuclear elements (LINEs) and short interspersed nuclear elements (SINEs) can integrate into the genome through the reverse transcription of an RNA intermediate (Kaessmann et al. 2009). Class II elements such as miniature inverted-repeat transposable elements (MITEs) move via a DNA intermediate by a cut-and-paste mechanism facilitated by the binding of transposase to terminal inverted repeats (TIRs) of the tranposable element (Craig et al. 2002). *EPSPS* gene amplification in glyphosate-resistant Palmer amaranth has been suggested to have evolved as a result of DNA transposon-mediated replication with possible involvement of MITEs flanking the *EPSPS* gene (Gaines et al. 2013). Although not naturally evolved, the role of *cis*-acting genetic elements such as amplification-promoting sequence (aps) of ribosomal DNA (rDNA) in the increased expression and copy number of acetolactate synthase (ALS) gene was reported in tobacco (*Nicotiana tabacum* L.) cell cultures (Borisjuk et al. 2000). This study suggests that the amplification of *ALS* gene results in an ALS-inhibitor-resistant phenotype. Transposable elements have been implicated in the mechanism of pesticide resistance mediated via cytochrome P450s. A MITE-like element was found in the upstream region of p450 gene *Cyp9m10* in pyrethroid-resistant *Culex quinquefasciatus* (Itokawa et al. 2010). Both class I and class II transposable elements were identified in P450 genes of corn earworm (*Helicoverpa zea*) (Chen and Li 2007). It would be interesting to look for such conserved genetic elements in the upstream or downstream regions of herbicide target genes in naturally evolved herbicide-resistant weed species.

Use of Molecular Cytogenetic Tools in Weed Science

In the past, cytological procedures were extensively used for structural and functional analyses of chromosomes, primarily to detect chromosomes during cell division. Later, banding techniques such as G-, R-, C-, and chromosomal nucleolar organizer regions (NORs) were developed for characterization of chromosomes (Kannan and Zilfalil 2009). With the advancement in molecular biology, use of cytogenetic tools have become even more valuable, especially to illustrate genome structure, genetic analyses and for the development of cytogenetic maps (Cui et al. 2015). High-resolution cytogenetic maps will help identify the precise location of genetic loci on the chromosomes. Such maps of individual chromosomes have been successfully constructed in many agronomically important crops, such as maize (Amarillo and Bass 2007), rice (Kao et al. 2006), Brassica (Howell et al. 2005; Xiong et al. 2010), cotton (Cui et al. 2015), and soybean (Walling et al. 2006). Despite the long history of cytology, use of high-resolution cytogenetic maps to understand chromosomal structure and/or function in weed species is limited.

Molecular cytogenetic techniques such as fluorescence *in situ* hybridization (FISH), and high resolution FISH on stretched DNA (fiber-FISH) have been developed to visualize individual genes and small DNA elements on chromosomes. Specifically, single to multiple-color probes, can be used for hybridization of labeled DNA fragments to intact chromosomes to detect the complementary sequences on the chromosomes (Tang et al. 2009; Fransz et al. 1996; de Jong et al. 1999). Furthermore, unlike the genetic linkage maps where the markers or DNA sequence is localized on a hypothetical chromosome, FISH offers a powerful and unique tool that allows the direct mapping of DNA probe on a chromosome (Wang et al. 2009). Thus, these high-resolution physical maps provide visible information regarding the position of DNA sequences including the distribution of repetitive sequences on the chromosomes (Schwarzacher 2003).

Genomic *in situ* hybridization (GISH) is also a widely used molecular cytogenetics technique for genome-specific chromosome painting in hybrids and polyploid species (Jiang and Gill 1994). It works on the principle of complementarity between DNA strands similar to FISH but unlike in

FISH where specific sequences are used as a DNA probe, in GISH, the whole genomic DNA from one species is used to target the genomic DNA of another. Several studies involving parental genome identification and foreign chromatin insertions for amphiploid species have been reported using GISH (Fedak et al. 2000; Chen et al. 1999). Disease resistance gene transfer between species can be identified using GISH (Cainong et al. 2015). Physical locations of particular genes of interest and their origin can be mapped using simultaneous FISH and GISH (Zheng et al. 2006).

Molecular cytogenetic tools, such as FISH, and/or fiber-FISH, essentially use DNA probes labeled with different colored fluorescent tags to visualize the amplified gene copies on specific regions of the genome. FISH can either be performed directly on metaphase chromosomes or interphase nuclei. On the other hand, fiber-FISH is a powerful tool that can be used for developing high-resolution physical maps. FISH was used to determine the location of the *EPSPS* gene on the chromosomes and its distribution in the genome of Palmer amaranth (Gaines et al. 2010). Using 1 kb cDNA fragment (1,044 bp) of the *EPSPS* gene as a DNA probe on metaphase and interphase nuclei of glyphosate-resistant Palmer amaranth, it was found that the amplified copies of *EPSPS* were randomly distributed throughout the genome. Similarly, FISH using the *EPSPS* probe (4,653 bp), identified a single and prominent hybridization site of the *EPSPS* gene localized on the distal end of metaphase chromosomes of glyphosate-resistant kochia, compared to a faint hybridization signal in glyphosate-susceptible samples. Further resolution of the hybridization signal using two *EPSPS* probes (~ 1,900 bp and ~ 2,500 bp) in fiber-FISH revealed tandem arrangement of ten *EPSPS* gene copies (one with an inverted *EPSPS* sequence) on an extended DNA fibers isolated from glyphosate-resistant kochia. The total length of the amplified *EPSPS* region was found to be 511 ± 26 kb, with a distance of 40–70 kb between two tandemly arranged *EPSPS* gene copies on the chromosome of glyphosate-resistant kochia (Jugulam et al. 2014). Thus, fiber-FISH accurately visualized the tandem organization of the *EPSPS* gene, which would otherwise be difficult to analyze using a Southern blot or quantitative PCR analysis. The application of molecular cytogenetic tools is not limited to capturing the location of the gene(s) of interest on the chromosomes. The scope of these techniques can be expanded to identify the successful introgression of herbicide-tolerant traits from a weed species to crops (Jugulam et al. 2015). In this study, transfer of dicamba (an auxinic herbicide) tolerance from wild mustard (*Sinapis arvensis*) into canola (*Brassica napus*) was achieved by a repeated backcross breeding. Importantly, stable introgression of the piece of DNA containing the dicamba tolerance gene from wild mustard into canola was demonstrated by FISH (Jugulam et al. 2015). Furthermore, molecular cytogenetic tools can be used to investigate functional and structural genomics and comparative evolutionary biology as well (Ananiev et al. 2009; Harper et al. 2012). Recent advancements in this discipline will improve our understanding of mechanisms that control the chromosomal behavior and dynamics at the gene level (Younis et al. 2015). High-resolution fiber-FISH can be applied to visualize epigenetic factors such as DNA methylation or histone modification as well (Koo et al. 2011).

Conclusions

In summary, applications of molecular cytogenetic techniques such as FISH and fiber-FISH have been valuable to understand the possible basis of *EPSPS* gene amplification resulting in evolution of glyphosate resistance in weed species. More importantly, although in many glyphosate-resistant weeds the *EPSPS* gene amplification has been reported, nonetheless, the mechanism of *EPSPS* gene duplication can be different in each species. Especially, marked differences have been seen in the distribution of *EPSPS* copies between glyphosate-resistant Palmer amaranth (Gaines et al. 2010) and kochia (Jugulam et al. 2014), suggesting that the basis of evolution of resistance via *EPSPS* gene amplification in response to glyphosate selection may be different in these two weed species. Future work using genomics and molecular cytogenetics will help understand the basis of *EPSPS* gene amplification in glyphosate-resistant weeds. Additionally, application of GISH combined with FISH in weed science can also assist in identification of the location of genes transferred among related weed species and also from weeds into crops or vice-versa.

LITERATURE CITED

Amarger V, Gauguier D, Yerle M, Apiou F, Pinton P, Giraudeau F, Monfouilloux S, Lathrop M, Dutrillaux B, Buard J and Vergnaud G (1998) Analysis of distribution in the human, pig, and rat genomes points toward a general subtelomeric origin of minisatellite structures. Genomics 52: 62–71.

Amarillo FI and Bass HW (2007) A transgenomic cytogenetic sorghum (*Sorghum propinquum*) bacterial artificial chromosome fluorescence *in situ* hybridization map of maize (*Zea mays* L.) pachytene chromosome 9, evidence for regions of genome hyperexpansion. Genetics 177: 1509–1526.

Ames D, Murphy N, Helentjaris T, Sun N and Chandler V (2008) Comparative analyses of human single and multilocus tandem repeats. Genetics 179: 1693–1704.

Ananiev EV, Wu C, Chamberlin MA, Svitashev S, Schwartz C, Gordon-Kamm W and Tingey S (2009) Artificial chromosome formation in maize (*Zea mays* L.). Chromosoma 118: 157–177.

Bass C and Field LM (2011) Gene amplification and insecticide resistance. Pest Manag. Sci. 67: 886–890.

Borisjuk N, Borisjuk L, Komarnytsky S, Timeva S, Hemleben V, Gleba Y and Raskin I (2000) Tobacco ribosomal DNA spacer element stimulates amplification and expression of heterologous genes. Nature Biotechnol. 18: 1303–1306.

Cainong J, Bockus WW, Feng Y, Chen P, Qi L, Sehgal SK, Danilova TV, Koo DH, Friebe B and Gill BS (2015) Chromosome engineering, mapping, and transferring of resistance to Fusarium head blight disease from *Elymus tsukushiensis* into wheat. Theor. Appl. Genet. 128: 1019–1027.

Chatham LA, Riggins CW, Martin JR, Kruger GR, Bradley KW, Peterson DE, Jugulam M and Tranel P (2015) A multi-state study of the association between glyphosate resistance and *EPSPS* gene amplification in waterhemp (*Amaranthus tuberculatus*). Weed Sci. 63: 569–577.

Chen J, Huang H, Zhang C, Wei S, Huang Z, Chen J and Wang X (2015) Mutations and amplification of *EPSPS* gene confer resistance to glyphosate in goosegrass (*Eleusine indica*). Planta 242: 859–868.

Chen Q, Conner RL, Laroche A, Ji WQ, Armstrong KC and Fedak G (1999) Genomic *in situ* hybridization analysis of *Thinopyrum* chromatin in a wheat-*Th. intermedium* partial amphiploid and six derived chromosome addition lines. Genome 42: 1217–1223.

Chen S and Li X (2007) Transposable elements are enriched within or in close proximity to xenobiotic-metabolizing cytochrome P450 genes. BMC Evol. Biol. 7: 46.

Craig NL, Craigie R, Gellert M and Lambowitz AM (2002) Mobile DNA II. Washington, DC: ASM Press.

Cui X, Liu F, Liu Y, Zhou Z, Zhao Y, Wang C, Wang X, Cai X, Wang Y, Meng F and Peng R (2015) Construction of cytogenetic map of *Gossypium herbaceum* chromosome 1 and its integration with genetic maps. Mol. Cytogenet. 8: 1–10.

De Jong JH, Fransz P and Zabel P (1999) High resolution FISH in plants–techniques and applications. Trends Plant Sci. 4: 258–263.

Devine MD and Eberlein CV (1997) Physiological, biochemical and molecular aspects of herbicide resistance based on altered target sites. pp. 159–185. *In*: Roe RM, Burton JD and Kuhr RJ (eds.). Herbicide Activity: Toxicology, Biochemistry and Molecular Biology. IOS Press, Amsterdam.

Devonshire AL and Field LM (1991) Gene amplification and insecticide resistance. Annu. Rev. Entomol. 36: 1–23.

Donn G, Tischer E, Smith JA and Goodman HM (1984) Herbicide-resistant alfalfa cells: an example of gene amplification in plants. J. Mol. Appl. Genet. 2: 621–635.

Eichler EE (2008) Copy number variation and human disease. Nat. Educ. 1: 1.

Fedak G, Chen Q, Conner RL, Laroche A, Petroski R and Armstrong KW (2000) Characterization of wheat-*Thinopyrum* partial amphiploids by meiotic analysis and genomic *in situ* hybridization. Genome 43: 712–719.

Fedoroff NV (2012) Transposable elements, epigenetics and genome evolution. Science 338: 758–767.

Field LM and Devonshire AL (1997) Structure and organization of amplicons containing the E4 esterase genes responsible for insecticide resistance in aphid *Myzus persicae* (Sulzer). Biochem. J. 322: 867–871.

Field LM, Devonshire AL and Forde BG (1988) Molecular evidence that insecticide resistance in peach-potato aphids (*Myzus persicae* Sulz.) results from amplification of an esterase gene. Biochem. J. 251: 309–312.

Flagel LE and Wendel JF (2009) Gene duplication and evolutionary novelty in plants. New Phytol. 183: 557–564.

Fransz PF, Alonso-Blanco C, Liharska TB, Peeters AJ, Zabel P and Jong JH (1996) High-resolution physical mapping in *Arabidopsis thaliana* and tomato by fluorescence *in situ* hybridization to extended DNA fibres. Plant J. 9: 421–30.

Gaines TA, Wright AA, Molin WT, Lorentz L, Riggins CW and Tranel PJ (2013) Identification of genetic elements associated with *EPSPS* gene amplification. PloS One 8: e65819.

Gaines TA, Zhang W, Wang D, Bukun B, Chisholm ST, Shaner DL, Nissen SJ, Patzoldt WL, Tranel PJ, Culpepper AS, Grey TL, Webster TM, Vencill WK, Sammons RD, Jiang JM, Preston C, Leach JE and Westra P (2010) Gene amplification confers glyphosate resistance in *Amaranthus palmeri*. Proc. Natl. Acad. Sci. USA 107: 1029–1034.

Harper LC, Sen TZ and Lawrence CJ (2012) Plant cytogenetics in genome databases. pp. 311–322. *In*: Harper LC (ed.). In Plant Cytogenetics. New York: Springer.

Heap IM (2016) International Survey of Herbicide Resistant Weeds. http://www.weedscience. org. Accessed December 31, 2015.

Howell EC, Armstrong SJ, Barker GC, Jones GH, King GJ, Ryder CD and Kearsey MJ (2005) Physical organization of the major duplication on *Brassica oleracea* chromosome O6 revealed through fluorescence *in situ* hybridization with *Arabidopsis* and *Brassica* BAC probes. Genome 48: 1093–1103.

Itokawa K, Komagata O, Kasai S, Okamura Y, Masada M and Tomita T (2010) Genomic structures of *Cyp9m10* in pyrethroid resistant and susceptible strains of *Culex quinquefasciatus*. Insect Biochem. Mol. Biol. 40: 631–640.

Jiang J and Gill BS (1994) Nonisotopic *in situ* hybridization and plant genome mapping: the first 10 years. Genome 37: 717–725.

Jugulam M, Niehues K, Godar AS, Koo DH, Danilova T, Bernd Friebe, Sehgal S, Varanasi VK, Wiersma A, Westra P, Stahlman PW and Gill BS (2014) Tandem amplification of a chromosomal segment harboring 5-enolpyruvylshikimate-3-phosphate synthase locus confers glyphosate resistance in *Kochia scoparia*. Plant Physiol. 166: 1200–1207.

Jugulam M, Ziauddin A, So KK, Chen S and Hall JC (2015) Transfer of dicamba tolerance from *Sinapis arvensis* to *Brassica napus* via embryo rescue and recurrent backcross breeding. PloS One 10: e0141418.

Kaessmann H, Vinckenbosch N and Long M (2009) RNA-based gene duplication: mechanistic and evolutionary insights. Nat. Rev. Genet. 10: 19–31.

Kannan TP and Zilfalil BA (2009) Cytogenetics: past, present and future. Malays J. Med. Sci. 16: 4–9.

Kao FI, Cheng YY, Chow TY, Chen HH, Liu SM, Cheng CH and Chung MC (2006) An integrated map of *Oryza sativa* L. chromosome 5. Theor. Appl. Genet. 112: 891–902.

Kejnovsky E, Hawkins JS and Feschotte C (2012) Plant transposable elements: Biology and evolution. pp. 17–34. *In*: Wendel JF, Greilhuber J, Dolezel J and Leitch IJ (eds.). Plant Genome Diversity: Plant Genomes, their Residents, and their Evolutionary Dynamics. Springer, New York.

Koo DH, Han F, Birchler JA and Jiang J (2011) Distinct DNA methylation patterns associated with active and inactive centromeres of maize B chromosomes. Genome Res. 21: 908–914.

Kubo K, Paape T, Hatakeyama M, Entani T, Takara A, Kajihara K, Tsukahara M, Shimizu-Inatsugi R, Shimizu KK and Takayama S (2015) Gene duplication and genetic exchange drive the evolution of S-RNase-based self-incompatibility in Petunia. Nat. Plants Doi:10.1038/nplants.2014.5.

Lisch D (2009) Epigenetic regulation of transposable elements in plants. Annu. Rev. Plant Biol. 60: 43–66.

Luo S, Peng J, Li K, Wang M and Kuang H (2011) Contrasting evolutionary patterns of the *Rp1* resistance gene family in different species of poaceae. Mol. Biol. Evol. 28: 313–325.

Ma Z and Michailides TJ (2005) Advances in understanding molecular mechanisms of fungicide resistance and molecular detection of resistant genotypes in phytopathogenic fungi. Crop Prot. 24: 853–863.

Malone JM, Morran S, Shirley N, Boutsalis P and Preston C (2016) *EPSPS* gene amplification in glyphosate-resistant *Bromus diandrus*. Pest Manag. Sci. 72: 81–88.

McClintock B (1951) Chromosome organization and genic expression. Cold Spring Harbor Symp. Quant. Biol. 16: 13–47.

Mouches C, Pasteur N, Berge JB, Hyrien O, Raymond M, de Saint Vincent BR, de Silvesti M and Georghiou GP (1986) Amplification of an esterase gene is responsible for insecticide resistance in a California Culex mosquito. Science 233: 778–780.

Muotri AR, Marchetto MCN, Coufal NG and Gage FH (2007) The necessary junk: new functions for transposable elements. Hum. Mol. Genet. 16: R159–R167.

Nagy ED and Bennetzen JL (2008) Pathogen corruption and site-directed recombination at a plant disease resistance gene cluster. Genome Res. 18: 1918–1923.

Ohno S (1970) Tandem duplication involving part of one linkage group at a time. pp. 89–97. *In*: Ohno S (ed.). Evolution by Gene Duplication. Springer, Berlin Heidelberg.

Paton MG, Karunarate SH, Gakoumaki E, Roberts N and Hemingway J (2000) Quantitative analysis of gene amplification in insecticide-resistant *Culex* mosquitoes. Biochem. J. 346: 17–24.

Perry GH, Dominy NJ, Claw KG, Lee AS, Fiegler H, Redon R, Werner J, Villanea FA, Mountain JL, Misra R, Carter NP, Lee C and Stone AC (2007) Diet and the evolution of human amylase gene copy number variation. Nature Genet. 39: 1256–1260.

Powles SB (2010) Gene amplification delivers glyphosate-resistant weed evolution. Proc. Natl. Acad. Sci. USA 107: 955–956.

Powles SB and Preston C (2006) Evolved glyphosate resistance in plants: biochemical and genetics basis of resistance. Weed Technol. 20: 282–289.

Powles SB and Yu Q (2010) Evolution in action: plants resistant to herbicides. Annu. Rev. Plant Biol. 61: 317–347.

Preston C and Wakelin AM (2008) Resistance to glyphosate from altered herbicide translocation patterns. Pest Manag. Sci. 64: 372–376.

Raymond M, Poulin E, Boiroux V, Dupont E and Pasteur N (1993) Stability of insecticide resistance due to amplification of esterase genes in *Culex pipiens*. Heredity 70: 301–307.

Robertson RR (2010) Physiological and biochemical characterization of glyphosate-resistant *Ambrosia trifida* L. Master's Thesis. West Lafayette, IN: Purdue University. 87 p.

Royle NJ, Clarkson RE, Wong Z and Jeffreys AJ (1988) Clustering of hypervariable minisatellites in the proterminal regions of human autosomes. Genomics 3: 352–360.

Salas RA, Dayan FE, Pan Z, Watson SB, Dickson JW, Scott RC and Burgos NR (2012) *EPSPS* gene amplification in glyphosate-resistant Italian ryegrass (*Lolium perenne* ssp. *multiflorum*) from Arkansas. Pest Manage. Sci. 68: 1223–1230.

Sammons RD and Gaines TA (2014) Glyphosate resistance: state of knowledge. Pest Manag. Sci. 70: 1367–1377.

Schoenfelder KP and Fox DT (2015) The expanding implications of polyploidy. J. Cell Biol. 209: 485–491.

Schwarzacher T (2003) DNA, chromosomes, and *in situ* hybridization. Genome 46: 953–962.

Segobye K (2013) Biology and ecology of glyphosate-resistant giant ragweed (*Ambrosia trifida* L.) Master's Thesis. West Lafayette, IN: Purdue University. 169 p.

Shaner DL, Lindenmeyer RB and Ostlie MH (2012) What have the mechanisms of resistance to glyphosate taught us? Pest Manag. Sci. 68: 3–9.

Slack A, Thornton PC, Magner DB, Rosenberg SM and Hastings PJ (2006) On the mechanism of gene amplification induced under stress in *Escherichia coli*. PLoS Genet. 2: e48.

Soukup SW (1974) Evolution by gene duplication. pp. 160. *In*: Ohno S (ed.). Springer-Verlag, New York. Teratol. 9: 250–251.

Steinrücken HC and Amrbein N (1980) The herbicide glyphosate is a potent inhibitor of 5-enolpyruvylshikimic acid-3-phosphate synthase. Biochem. Biophs. Res. Commun. 94: 1207–1212.

Tang X, De Boer JM, Van Eck HJ, Bachem C, Visser RG and De Jong H (2009) Assignment of genetic linkage maps to diploid *Solanum tuberosum* pachytene chromosomes by BAC-FISH technology. Chromosome Res. 17: 899–915.

Taylor JS and Raes J (2004) Duplication and divergence: the evolution of new genes and old ideas. Annu. Rev. Genet. 38: 615–643.

Van der Hoorn RA, Kruijt M, Roth R, Brandwagt BF, Joosten MH and De Wit PJ (2001) Intragenic recombination generated two distinct *Cf* genes that mediate AVR9 recognition in the natural population of *Lycopersicon pimpinellifolium*. Proc. Natl. Acad. Sci. USA 98: 10493–10498.

Wagner GP, Pavlicev M and Cheverud JM (2007) The road to modularity. Nat. Rev. Genet. 8: 921–931.

Walling JG, Shoemaker R, Young N, Mudge J and Jackson S (2006) Chromosome-level homeology in paleopolyploid soybean (*Glycine max*) revealed through integration of genetic and chromosome maps. Genetics 172: 1893–1900.

Wang K, Yang Z, Shu C, Hu J, Lin Q, Zhang W, Guo W and Zhang T (2009) Higher axial-resolution and sensitivity pachytene fluorescence *in situ* hybridization protocol in tetraploid cotton. Chromosome Res. 17: 1041–1050.

Wiersma AT, Gaines TA, Preston C, Hamilton JP, Giacomini D, Buell CR, Leach JE and
    Westra P (2015) Gene amplification of 5-enol-pyruvylshikimate-3-phosphate synthase
    in glyphosate-resistant *Kochia scoparia*. Planta 241: 463–474.
Xiong Z, Kim JS and Pires JC (2010) Integration of genetic, physical, and cytogenetic maps for
    Brassica rapa chromosome A7. Cytogenet. Genome Res. 129: 190–198.
Younis A, Ramzan F, Hwang YJ and Lim KB (2015) FISH and GISH: molecular cytogenetic
    tools and their applications in ornamental plants. Plant Cell Rep. 34: 1477–1488.
Yu Q, Jalaludin A, Han H, Chen M, Sammons RD and Powles SB (2015) Evolution of a double
    amino acid substitution in the 5-enolpyruvylshikimate-3-phosphate synthase in *Eleusine
    indica* conferring high-level glyphosate resistance. Plant Physiol. 167: 1440–1447.
Zhang J (2003) Evolution by gene duplication: an update. Trends Ecol. Evol. 18: 292–298.
Zheng Q, Li B, Mu S, Zhou H and Li Z (2006) Physical mapping of the blue-grained gene(s)
    from *Thinopyrum ponticum* by GISH and FISH in a set of translocation lines with different
    seed colors in wheat. Genome 49: 1109–1114.

Chapter 11

Applications of Genomics in Weed Science

Todd A. Gaines,[1] Patrick J. Tranel,[2] Margaret B. Fleming,[1] Eric L. Patterson,[1] Anita Küpper,[1] Karl Ravet,[1] Darci A. Giacomini,[2] Susana Gonzalez[3] and Roland Beffa[3]

Introduction

The information that determines the structure and function of an organism is contained by its DNA. This genetic information guides the production of specific RNA molecules, which in turn guide the production of specific protein molecules. As elucidated by Watson and Crick (1953), genetic information is maintained by a double helix structure, in which two chains having sugar/phosphate backbones are held together by hydrogen-bond pairing of adenine (A) with thymine (T) and guanine (G) with cytosine (C). The specific arrangement of the A, T, G, and C nucleotide bases holds the information that defines an organism.

At one time, determining the specific order of all the nucleotides of a single gene, which might consist of a few hundred nucleotides—let alone of a whole genome, which might consist of a few hundred million nucleotides—was a daunting task. Sanger et al. (1977) introduced a method for DNA sequencing (a variation of which is still used today) that opened

[1] Department of Bioagricultural Sciences and Pest Management, Colorado State University, Fort Collins, CO, USA.
[2] Department of Crop Sciences, University of Illinois, Urbana, IL, USA.
[3] Weed Resistance Competence Center, Bayer Crop Science, Frankfurt am Main, Germany.

the doors for whole-gene sequencing and even sequencing of viral genomes (Hutchison 2007). Advances in, and automation of, Sanger sequencing technology enabled more ambitious gene sequencing projects, such as the sequencing of the first entire eukaryotic (yeast) genome in 1996 (Goffeau et al. 1996). Many additional genomes were sequenced during the following ten years. However, due to the high cost, genome sequencing efforts were limited to organisms (such as viruses and prokaryotes) that have very small genomes, model organisms [e.g., *Arabidopsis thaliana*; The *Arabidopsis* Genome Initiative (2000)], economically important species [e.g., rice (*Oryza sativa*) (Yu et al. 2002) and mosquito (*Anopheles gambiae*) (Holt et al. 2002)] and humans (Lander et al. 2001). In 2005, however, DNA sequencing was revolutionized by the introduction and commercialization of massively parallel sequencing technology (454 pyrosequencing) (Margulies et al. 2005; Rothberg and Leamon 2008). This began the era of "next-generation" sequencing, in which 454 pyrosequencing and other new sequencing technologies caused DNA sequencing costs to plummet. From 2007 to 2015, the cost to generate 3 billion nucleotides of sequence data (equivalent to the human genome) went from about $10 million to less than $10 thousand (Wetterstrand 2015). Whole-genome sequencing projects can now be considered for essentially any organism.

To be sure, genome sequencing is still far from trivial. Although generation of sequence data is no longer the major hurdle, assembly of the data remains a bioinformatics challenge, since the most accurate and inexpensive DNA sequencing technologies generate relatively short reads. Furthermore, many non-model organisms—including many weed species – present additional challenges to whole genome sequencing and assembly, such as high heterozygosity and polyploidy. And finally, extracting biological meaning from the mountains of data requires not only having defined research questions and expertise in bioinformatics to mine the data for answers, but also often requires downstream experimentation to test hypotheses that are generated.

Despite the remaining challenges to genomics, and the promise of even further advances in the technology, the era of weed genomics is upon us. Already the first draft genome of an economically important weed species, horseweed (*Conyza canadensis*), has been published (Peng et al. 2014b). In this chapter, we will begin with an overview of plant genomes already assembled, and then discuss methods, applications, and opportunities for weed genome assembly. Finally, we will discuss transcriptomics approaches (which focuses on expressed genes within the genome) in weed science that are already making important contributions to our understanding of weed biology.

Assembled Plant Genomes

Assembly of a plant genome requires overcoming some significant challenges. The first major challenge arises from the presence of the two non-nuclear genomes belonging to the chloroplast and the mitochondria. The sequences belonging to these genomes must be removed either prior to sequencing through clever DNA extraction or after sequencing using bioinformatics. The second challenge is a consequence of the ability of higher plants to tolerate a large degree of genome remodeling and rearrangement. Plants consequently have "messy" and/or extraordinarily large genomes compared to those of other eukaryotes, making sequencing of these genomes impossible until recently, with the development of cheaper sequencing protocols and programs to handle unwieldy data sets (see Fig. 1 for sizes of published plant genomes). Finally, DNA sequences can be repeated, either in tandem or scattered throughout the genome, as a result of ancient polyploidizations, movement of transposable elements, or non-homologous recombination during meiosis. These duplications can range from single base pairs up to entire chromosomes. The most extreme form of sequence duplication is polyploidy, which means the genome contains more than the normal two complete sets of chromosomes found in diploids. Many important crop species have very high genomic complexity, often due to their polyploid genomes (e.g., bread wheat, a hexaploid organism), or due to the significant amounts of non-coding DNA in their genome (e.g., corn, a genome containing ~ 90% transposable elements). Determining the location of repeated sequences in the complete genome, their number, and their arrangement can sometimes be a nearly insurmountable task.

Despite these complexities, researchers have assembled many plant genomes. The first plant genome (and third eukaryotic genome) to be assembled was that of *Arabidopsis thaliana*, which required four years of dedicated effort and the international collaboration of over 150 scientists through the Arabidopsis Genome Initiative (2000). The release of the genome was announced in a Nature paper and was rightly considered a major accomplishment. The next plant genomes published were those of two related rice varieties, *Oryza sativa* ssp. *indica* (Yu et al. 2002) and ssp. *japonica* (Goff et al. 2002). Several algae genomes were then sequenced and assembled, and the next true plant genome was not published until 2006, that of *Populus trichocarpa* (Tuskan et al. 2006). Since 2006, the number of plant genome assemblies has increased dramatically, leading to the publication of over 30 genomes in 2013 (Michael and Jackson 2013). With the advent of lower-cost next-generation sequencing, as well as the accumulation of over fifty complete plant genome assemblies, assembling a new genome has become a project almost every molecular biology lab can consider.

Recently, several extremely challenging genomes have been assembled. A draft genome of bread wheat, which has a large (17 Gb) hexaploid

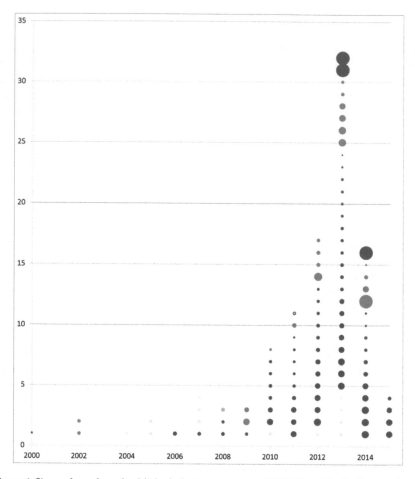

Figure 1. Size and number of published plant genomes, as of 2015. Size of dot indicates relative haploid genome size, from 100 Mb (smallest) to 20 Gb (largest) (not to scale). Blue, angiosperm monocots; yellow, algae; red, angiosperm dicots; purple, gymnosperms; green, bryophytes; orange, lycopodiopytes. Data from NCBI Genome (http://www.ncbi.nlm.nih.gov/genome/) and JGI Genome Portal (http://genome.jgi.doe.gov/viridiplantae/viridiplantae.info.html).

genome, was published in 2012 (Brenchley et al. 2012). Several gymnosperm genomes, each containing over 20 Gb, have been recently published— two in 2013, *Picea abies* (Nystedt et al. 2013) and *Picea glauca* (Birol et al. 2013), and one in 2014, *Pinus taeda* (Zimin et al. 2014). The assembly of these genomes required developing new software and hardware capabilities for data analysis (Neale et al. 2014), since the sheer scale of the genome assembly would overwhelm standard assembly and computing platforms.

In several cases, complex genome assemblies, particularly from polyploid organisms, were simplified by sequencing DNA from a plant in a unique genetic state, generally the haploid generation. Haploid sequencing reduces heterogeneity during sequencing, because only one copy of the genome is present and therefore there is no heterozygosity. Potato, which is an autotetraploid, was assembled from both a doubled monoploid and a haploid source (Xu et al. 2011). Muskmelon (*Cucumis melo*) was sequenced from a doubled haploid (Garcia-Mas et al. 2012), as was banana (*Musa acuminata*) (D'Hont et al. 2012). The haploid form of *Citrus clementina* (Wu et al. 2014) and the haploid megagametophyte of *Pinus taeda* (Zimin et al. 2014) were sequenced.

Another useful technique for simplifying genome assembly is to scaffold the new genome assembly on that of a close relative, as in the case of crucifer relatives of *A. thaliana*: *Leavenworthia alabamica*, *Sisymbrium irio*, and *Aethionema arabicum*. The assembly of these crucifers was facilitated by the ability to compare the genome structure and conserved elements with that of the already sequenced near relative *A. thaliana* as well as five other crucifer species (Haudry et al. 2013). This comparison is possible because related species result from divergence of a common ancestral genome. If the amount of divergence time is low (as in the case of the crucifers), significant similarities in sequence and structure remain, which can be used to scaffold an assembly. As of September 2015, whole-genome sequencing of over 1000 Arabidopsis accessions has been completed, permitting detailed analysis of the effect of genotype on every level of plant physiology (information available at 1001genomes.org; a publication is forthcoming).

The majority of published plant genomes have been either of model species (*A. thaliana*, *Brachypodium distachyon*, *Petunia*) or of crop species and their near relatives. Plant genomes that are interesting for taxonomic and systematics studies are just beginning to be considered. Only one bryophyte genome, *Physcomitrella patens* (Rensing et al. 2008) (a model species) and one lycopodiophyte genome (*Selaginella moellendorffii*) (Banks et al. 2011) have been assembled, as well as *Amborella trichopoda* (Albert et al. 2013), which is thought to be the descendent of the root of the angiosperm branch. However, now that genomic sequencing as well as assembly have become more routine, scientists are branching out, assembling the genomes of plants from diverse clades.

Genomics for Weedy Species

Methods in genome sequencing and assembly

Genomic sequencing is deciphering the exact order of nucleotides in a segment of DNA. Molecular biologists, chemists, engineers, and computer scientists have collaborated to develop many sequencing technologies that are commonly referred to as platforms. In this section, we will discuss several of these platforms, the types of data they produce, and how the data can be used to enable genomics.

Sanger sequencing

DNA sequencing began with the work of Frederick Sanger in the 1970s using DNA elongation inhibitors (Sanger et al. 1977). His technique is now commonly referred to as Sanger sequencing. In Sanger sequencing, DNA is replicated using the same cycling conditions as polymerase chain reaction, but rather than including only deoxynucleotides (dNTPs: ATP, TTP, GTP, and CTP), a small percentage of labeled dideoxynucleotides (ddNTPs) are also added to the reaction. When the polymerase enzyme inserts a nucleotide during nucleotide extension, it has a chance of incorporating a ddNTP rather than a dNTP. This incorporation terminates any further elongation and tags the end of the sequence with the label that is associated with the incorporated ddNTP. Since the incorporation of a ddNTP is random, some DNA synthesis reactions will terminate very early and make short fragments, while others will carry on longer before termination occurs. In this way, it is possible to generate a pool of nucleotide fragments of varying length, each tagged with a label that represents one of the four bases. These fragments are then separated based on their size using capillary gel electrophoresis. As the fragments reach the end of the gel, the fluorescent label corresponding to the terminating ddNTP is detected.

Sanger sequencing is relatively inexpensive, extremely accurate, and very accessible. However, Sanger sequencing requires a primer to begin the sequencing reaction, which in turn requires some knowledge of the nucleic acids that are going to be sequenced. Sanger sequencing also only gives one read at a time with a maximum length of approximately 850 bp. To generate long continuous pieces of sequence, multiple reactions need to be performed and the resulting sequences aligned.

Next-generation sequencing

New sequencing platforms take advantage of the miniaturization of computers and sensors, as well as increases in data storage and enzyme engineering. Many of these "next-generation sequencing" (NGS) platforms share similar features. The general procedure begins by fractionating a pool of nucleic acids (DNA or RNA) into small pieces of a desired length. The nucleic acids can be derived from any source, including whole genomes. These pieces are next ligated to an oligonucleotide adapter that can be used to (A) identify the sample origin and (B) initiate the sequencing reaction by providing a site from which polymerase extension can begin. The tagged fragments are then separated by attaching them to a solid surface, bead, or well. The fragment is amplified by PCR (using the adaptor sequence as a primer) and light signals are detected from each added base to determine the sequence.

NGS technologies also have advantages and disadvantages. The primary advantage is that NGS allows for a huge amount of sequence data to be generated for a relatively small amount of time, space and money. However, sequence data generated by NGS is fragmented and has no context, meaning that each sequence is not identifiable until it is properly aligned. Additionally, sequences are usually relatively short (< 1000 bp). Therefore, the sequences need to be combined together into long contiguous stretches of sequence known as contigs in a process called assembly, which generally requires large amounts of computational power. We will briefly cover the more common platforms used to make genomes and transcriptomes. New NGS platforms are continually introduced to the market, making this a rapidly developing field of study.

Pyrosequencing

The first NGS platform invented was pyrosequencing (Ronaghi et al. 1996). In pyrosequencing, fractionated nucleic acids are attached to small beads with oligonucleotide complementation. These beads are inserted into wells that are almost the exact size of the beads so that two beads cannot fit inside the same well. The nucleotides are amplified in the wells using PCR until a population of identical sequences exists in each well. Another round of PCR is performed using a polymerase coupled with a chemiluminescent protein that only emits light when the polymerase adds a nucleotide. Each of the four bases is washed over the beads sequentially. If there is a flash of light in the well, then a nucleotide has been added. If not, then that nucleotide was not added. The sequencer automatically washes adenosine, tyrosine, cytosine and guanine in rotation until all reactions terminate (Ronaghi et al. 1998). Each well is individually monitored and the luminescence data is correlated to the addition of the nucleotides. These wells are extremely small, with hundreds of thousands fitting onto an area no larger than a stamp. Pyrosequencing generally has a low error rate and sequence reads from 300 to 700 bases, and produces in the range of 1 million reads per sample.

Illumina

In Illumina sequencing, short fragments of DNA are tagged with adaptors and ligated to a glass slide covered in short oligonucleotides complementary to the adaptors on the fragments (Mayer et al. 2011). After the fragments are bound, they are amplified using PCR, creating a small region of the slide that is populated with the exact same sequence. The slide is then washed in the four nucleic acids, each one tagged with a different fluorophore. All four bases are washed over the slide, their fluorophore detected and then washed away. The polymerase enzyme can only add one base at a time and so the process repeats for 100–250 cycles, each cycle adding a single base. Every cluster of identical oligonucleotides are read simultaneously, with millions of sequences being generated on a single slide. Both ends of a fragment can be sequenced, creating paired-end reads.

PacBio

PacBio sequencing is the first of so-called "third generation sequencing" technologies. PacBio, also known as single molecule real time (SMRT) sequencing, is unique from the second-generation technologies because it does not require an initial amplification step (Eid et al. 2009; Levene et al. 2003). As the name SMRT suggests, single DNA molecules (not pools or clusters) are sequenced. A DNA polymerase is fixed to the bottom of an extremely small well, formed out of aluminum (Korlach et al. 2008). This well has special optical properties that direct any photons released from a fluorophore directly down into a sensor. The nucleic acids to be sequenced are fractionated into large fragments (> 5000 bp) and circularized. These circular pieces of DNA are washed over the wells and the polymerases at the bottom of the wells each associate with a single circularized nucleic acid molecule. A single-cycle PCR is performed using fluorescent dNTPs that emit photons as they are incorporated in the PCR product; each dNTP emits photons of a different wavelength.

PacBio has several advantages to second-generation sequencing. First, the entire reaction only takes a few hours because individual dNTPs do not need to be added sequentially and washed off. Second, the sequence generated can be very long (> 20,000 bp). This can improve downstream sequence assembly. Third, the circularized piece of DNA can be read several times in each well, generating higher coverage for each genomic region. PacBio currently still needs to be combined with data from other platforms for *de novo* assembly because it has a high error rate. Currently, error rates fall anywhere between 10 to 20% of base pairs miscalled, and several insertions and deletions can be erroneously added to the sequence (Quail et al. 2012). Typically PacBio data is error corrected with more accurate sequencing methods in a hybrid-platform assembly.

Assembly

The greatest advantage of NGS platforms is the large amount of sequence data that can be generated in a short amount of time. Unfortunately, this can also be one of the greatest disadvantages, because all of these short reads are created with no guide as to where they belong. In a chromosome of 1,000,000 base pairs it would take 10,000 non-overlapping 100 bp sequences to span the entire length. Given that most genomes are several hundred million to several billion base pairs in length, it takes far more data to even begin to know the sequence of an entire genome. Additionally, it is generally not sufficient for each nucleotide to be sequenced only once due to possible errors in sequencing. Finally, once a sufficient amount of redundant sequence has been obtained, the sequences must be assembled into contigs, with the largest possible contig being a single chromosome. This often means the final number of reads to be assembled into contigs is in the millions, making assembly a highly computationally intensive task. Commonly used assemblers for NGS data include SOAPdenovo (Luo et al. 2012), Oases (Schulz et al. 2012), Celera (Myers et al. 2000), Newbler (Roche Diagnostics Corporation), Trinity (Grabherr et al. 2011), and ALLPATHS (Butler et al. 2008).

The first step in assembly is quality control of the read data. All sequencing platforms have at least some fraction of nucleotides that have been incorrectly identified, resulting in a characteristic error rate. Statistical values can be assigned to every nucleotide, and this value represents the confidence value. During the quality control process, some or all bases from reads can be removed based on their confidence value to ensure all sequences used for assembly meet defined quality criteria. Additionally, any adaptor sequence needs to be removed.

After quality control, the data move to an alignment program (algorithm) that takes the reads and aligns them based on their nucleotide order. All reads are compared to other reads and the program begins to build continuous stretches of sequence known as contigs. Some reads will be redundant to the sequence of the contigs and will not be useful in extending the contig sequence, but they can be useful for double checking the already assembled contig. Reads that partially overlap with the growing ends of the contig can be incorporated to extend the contig. The extension of contigs is typically done one or a few base pairs at a time, with a majority of the read length aligning to the end of the existing contig and only a small fraction of overhang incorporated into the contig. The more nucleotides the read overlaps the contig and the closer the similarity of the overlap, the more confidence the assembly program has to incorporate that read into the assembly.

This assembly method has major drawbacks. One of the largest problems is that single nucleotide polymorphisms (SNPs), insertions, and deletions in the reads are difficult to handle and may halt contig extension. Another possibility is that several different contigs may be assembled despite only minor differences between them. In reality these contigs belong to only one location in the genome, but it appears in the assembly as if there are several regions.

Another problem comes when there are redundant or repeated regions of the genome. When these regions are sequenced they generate the same reads, with the same sequence. These redundant reads align to each other and are collapsed into a single contig, when in reality that sequence is located in several places. In these two ways, it is possible to get several contigs that represent the same location, or a single contig that represents several locations.

These problems are not easily solved. To overcome the problem of polymorphisms, it is important to sequence individuals with low heterozygosity and variability in their genomes. In plants, this is typically done by inbreeding the plants to be sequenced. This can be especially difficult in species that only outcross. Additionally, parameters in the assembly program may be manipulated to allow for some amount of heterozygosity. However, this usually reduces confidence in the assembled contigs and reduces the quality of the assembly. Redundancy can be overcome by generating long reads that span the length of the repeat, with the ends of the reads in unique sequences. The longer the reads, the easier it is to assemble redundant regions into separate contigs. Mate pair reads are also a common method used to generate scaffolds bridging repetitive regions. Mate pair libraries are generated by selecting larger fragments (e.g., 5 or 10 kb) for paired-end sequencing.

Applications—Population Genomics

Whole genome sequencing techniques

Though the cost of DNA sequencing has fallen rapidly since the introduction of NGS platforms, whole-genome sequencing of hundreds of samples is still expensive. In certain research areas like population genetics and breeding, which deal with large numbers of individuals or lines, whole genome sequencing is not cost-effective and often unnecessary. Analysis of just a subsection of the genome is a cost-competitive alternative. This technique identifies large numbers of polymorphisms, used as genetic markers for gene/quantitative trait locus (QTL) mapping, genetic linkage analysis, germplasm characterization, molecular marker discovery, construction of high-density genome and haplotype maps, identification of candidate genes, within-species diversity studies, and genome-wide association studies (GWAS) (Davey et al. 2011; Elshire et al. 2011). The use of reduced libraries is especially interesting in the case of highly repetitive content, polyploidy, and presence or absence of homeologs (Deschamps et al. 2012). Compared to whole genome sequencing and traditional marker development (single locus, PCR-based markers, e.g., microsatellites), using NGS for marker discovery identifies polymorphisms more cost-effectively and faster due to reduced sample handling, PCR, and purification steps (Davey et al. 2011; Deschamps et al. 2012; Elshire et al. 2011).

Depending on the availability of a reference genome and the importance of marker accuracy and genome coverage, different genetic marker technologies using NGS are available. They range from reduced-representation sequencing, such as reduced-representation libraries (RRLs) and complexity reduction of polymorphic sequences (CRoPS), to restriction-site-associated DNA sequencing (RAD-seq) and low coverage genotyping [e.g., multiplexed shotgun genotyping (MSG) and genotyping by sequencing (GBS)]. In all of these methods, DNA is digested and a selected subset of sequence fragments is sequenced on an NGS platform. The methods differ by fragment size selection and shearing, as well as the use of barcodes and PCR steps (Davey et al. 2011).

In the case of GBS, DNA is extracted from plant tissue and quantified. To reduce the genome complexity, the DNA is digested with a restriction enzyme (*Ape*KI or *Pst*I/*Msp*I) (Elshire et al. 2011; Poland et al. 2012) creating fragments of different sizes. Then, forward and reverse adaptors and unique barcodes are ligated to the ends of the fragments to be able to identify the different samples after pooling them into a single tube. Subsequently, the condensed DNA samples are amplified by PCR and the DNA fragments within the desired size range of 100–400 bp are multiplex sequenced on an NGS platform (e.g., Illumina). This reduces the amount of DNA reads to a small fraction and allows for partial but genome-wide coverage (Davey et al. 2011; Sonah et al. 2013). The resulting files from the sequence reads are then sorted by their barcode to generate a sequencing file for each sample separately. If a reference genome is available, the reads are aligned and SNPs are called. Otherwise, *de novo* assembly [e.g., via the Universal Network Enabled Analysis Kit (UNEAK) (Lu et al. 2013)] is performed. During subsequent bioinformatics analysis, the hundreds of thousands of called SNPs are scored depending on alignment confidence and read quality (Elshire et al. 2011; Glaubitz et al. 2014). High confidence for SNP calls is important because NGS technologies generate relatively short reads and GBS relies on low sequencing coverage.

Applications in plant breeding

GBS is a particularly valuable resource in the field of plant breeding for genome- or population-wide studies. As an example, in GWAS the DNA of many individuals from many populations with and without a desired phenotype is compared. Differences in SNPs can help associate alleles with the trait of interest. This is especially helpful with quantitative trait loci (QTL) since certain traits, such as yield, result from combinatorial effects (Deschamps et al. 2012). To verify such an association with enough data, hundreds of thousands of markers are required. GBS can also be useful in the construction of genetic linkage maps to locate genes of interest on a chromosome. Furthermore, it is a powerful tool for the discovery of genome-wide molecular markers for marker-assisted selection (MAS), in which individuals are selected based on the presence/absence of specific markers rather than on phenotypic traits. MAS is a helpful technique in cases where the trait exhibits low heritability, or is too costly or slow to measure. Genomic selection (GS) goes even further and combines molecular markers with information on phenotype and pedigree to increase breeding accuracy (He et al. 2014). Another application of GBS is genetic diversity studies. Polymorphisms found through GBS can be used to score variations among individuals of a population or among populations by looking at recombination breakpoints. This allows one to infer genetic relationships in regard to ploidy, ecotype, and geographical distribution. It also enables the construction of genetic maps and phylogeographic trees based on the probabilities of recombination (Baird et al. 2008; Deschamps et al. 2012).

Applications in weed science

GBS also provides a valuable resource to address research questions in the field of weed science. As an example, Schneeberger et al. (2009) were able to narrow down the candidate region for a mutation in Arabidopsis plants which caused slow growth and light green leaves. They did so by genome-reduction sequencing a pool of 500 F_2 plants, a number of individuals that would have been prohibitively expensive to fully sequence. When a reference genome is available, this technique offers a possibility to find candidate regions for mutations that could confer herbicide resistance.

Similar to studies in plant breeding, GBS could help find genes and QTL associated with phenotypes that make weeds successful, such as elevated seed production, extended seed dormancy, thick cuticle layers, fast growth, and adaptability to abiotic stresses such as drought and high salinity. Identifying the genes or QTL responsible for these traits could open new ways to genetically modify crops for better survival and higher yield (Maughan et al. 2011).

Another important GBS application is determining genetic relatedness and gene flow in weeds. Obtaining SNP genotypes from thousands of loci can replace earlier marker systems such as AFLPs and SSRs, enabling the study of many plants from different locations to reveal migration patterns of individuals and certain traits. For example, the evolution and dispersion of interspecific hybrids of invasive weeds can be investigated as seen with hybrids of diffuse and spotted knapweed (Blair and Hufbauer 2010). In this case, results from amplified fragment length polymorphism (AFLP) markers suggested that the hybrid was most likely introduced together with diffuse knapweed in the early invasion of North America instead of occurring after introduction. Burrell et al. (2015) used GBS for a study on the origins and invasion history of *Imperata cylindrica*, increasing the number of informative polymorphisms exponentially over earlier marker systems. GBS can be used to follow crop introgressions into wild and weedy populations. Burgos et al. (2014) used GBS to discover that the introgression of ALS resistance from Clearfield rice had changed the genetic structure of US weedy rice populations. Okada et al. (2013) followed the evolution and spread of glyphosate resistance in *Conyza canadensis* in California using SSR markers. Based on the genetic diversity and population structure it was shown that the multiple independent origins of resistance had occurred within the sampling area and then resistance expanded due to positive selection with the herbicide glyphosate. Furthermore, the results showed that resistance was present several years before it was first reported in the area. Studies such as these allow researchers to follow gene flow via wind dispersal of pollen and seed movement. This information is of great importance for the development of successful weed and weed resistance management strategies.

It is expected that GBS will become even faster, higher in quality, and increasingly affordable in the near future. Therefore, it provides a new avenue in the identification and mapping of markers. Also, since it does not require a reference genome, it is of value especially in the field of weed science since most weed species have not yet been fully sequenced. Population genomics in weedy species is predicted to advance rapidly with the further development of already powerful techniques like GBS.

Applications of Genomics for Weedy Species

Genomics approaches in weedy species are beginning to advance beyond studies of the sequence and expression of single genes, but overall genomics resources for weedy species remain limited. The first genomic sequence reported was for waterhemp (*Amaranthus tuberculatus*) (Lee et al. 2009). This study used the 454 pyrosequencing technology to sample the waterhemp genome. Although the depth of sequence coverage was not sufficient to assemble the complete genome, partial sequence of several herbicide target-site genes were obtained, marking a major point of progress in obtaining genomic sequence resources for weeds.

A draft genome of horseweed (*Conyza canadensis*) was assembled using multiple sequencing platforms including 454, Illumina HiSeq (paired-end reads and mate-pair libraries), and Pac-Bio (Peng et al. 2014b). This draft genome contained 20,075 contigs of N50 20,764 bp, and 13,966 scaffolds of N50 33,561 bp. The longest assembled contig was 102,072 bp. The total assembly was 311.27 Mb, representing 92.3% coverage of the 335 Mb genome, including the complete chloroplast genome and a nearly complete mitochondrial genome. Sequence-based annotation identified 44,592 predicted protein-coding genes in the assembly. A notable result was an apparent size increase, relative to Arabidopsis, of gene families implicated in herbicide detoxification, including cytochrome P450s, glutathione-S-transferases, ATP-binding cassette transporters, and glycosyltransferases. Such increase in variation of important gene families may be a genetic trait underlying weediness (Peng et al. 2014b). Future genome assemblies for weedy species will likely utilize a similar approach of combining multiple NGS platforms of high coverage with short length, and low coverage with longer length, to obtain both longer and more accurate assemblies.

The genome of *Lolium perenne*, a forage species with weedy relatives, is currently being sequenced and assembled. Comparative genomics using the published barley genome has enabled the available *L. perenne* sequence to be utilized for genetic mapping and synteny studies (Pfeifer et al. 2013). *L. perenne* is very closely related to *Lolium multiflorum* and *Lolium rigidum*, both weedy species with the same number of chromosomes. Genomic studies of weedy *Lolium* species will benefit from comparison to the *L. perenne* genome.

NGS and genomics approaches have applications for characterizing known resistance-endowing mutations such as target-site mutations in a large number of individuals. Work recently completed in *Poa annua* used Illumina sequencing from mRNA to assemble the two homeologs of the acetolactate synthase gene in this tetraploid species (Chen et al. 2014). Homeolog-specific PCR primers were then designed to amplify the two homeologs from genomic DNA for sequencing. Such an approach can be utilized to amplify multiple herbicide target-site genes for high-throughput sequencing on platforms such as Illumina. This enables a large number of samples and genome reduction approaches (such as first amplifying the desired target-site sequence) to multi-plex 100 or more samples and take advantage of the high number of sequences generated on NGS platforms.

Transcriptomics for Weedy Species

Methods

Transcriptomics: from a single known gene to multiple unknown genes

Transcriptomics is the study of the transcriptome, the complete set of RNA transcripts that are produced by an organism under specific circumstances. Although study of transcript accumulation has become a usual laboratory practice after the first genomes were partially sequenced, pioneer techniques such as northern blots, real-time RT-PCR or microarrays rely on the hybridization of a specific probe and therefore require a perfect knowledge of the target RNA sequence (Fig. 2). This mostly restricts their use to well-characterized model organisms. Another caveat is the coverage of the

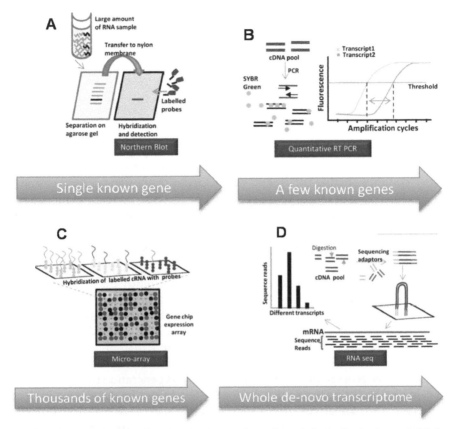

Figure 2. Progress in methods for expression analysis of genes from a few genes to multiple ones to whole transcriptome analysis. (**A**) Northern blotting is used for the detection of a single/few genes by hybridization of labeled probes to localized RNA samples run on an agarose gel and transferred to a membrane. (**B**) Quantitative PCR is a real time estimation of the levels of transcript of one/few genes, in the cDNA samples reverse transcribed from RNA, often calculated as fold changes with respect to a control gene. (**C**) In microarray, cRNA hybridized to probes spotted on a chip gives a signal value, which can be used to detect the differentially expressed as well as specific known genes. (**D**) RNA sequencing is transcript sequencing by NGS and shows the transcript number of known as well as novel genes along with alternative splicing and transcript fusions. Adapted from Agarwal et al. (2014).

transcript analysis. Northern blots and real-time RT-PCR analyze up to a few transcripts, while microarrays developed from sequenced genomes allow hybridization of dozens to thousands of genes at once. Microarrays paved the path to the most recent technologies, based on NGS, to allow high-throughput genome-wide transcriptome sequencing and detection of the complete set of RNA for both model and non-model organisms (Fig. 2).

NGS-based transcriptome approaches yield quantitative data (transcript expression level) and qualitative data (transcript sequences) without the need for pre-existing genomic resources. This latter point made RNA-sequencing methods (RNA-Seq) a revolutionary tool for molecular research once the target sequences go beyond known genomic sequences.

Transcriptomics: potential applications

Transcriptomics can have a wide array of potential applications, depending on plant species of interest and the research focus:

- For plant genomes with barriers to genome sequencing, transcriptome sequences provide a rapid and inexpensive method to access the "coding potential" of diverse plant species. Following *de novo* genome sequencing, transcriptome sequencing can also provide evidence for genome assembly accuracy and completeness, and can validate gene model predictions, splice variants and polymorphism. Therefore, RNA-Seq has become a proxy or a complementary approach for *de novo* genome sequencing.

- For gene expression analysis, RNA-Seq is currently the most powerful approach to identify genes differentially expressed among samples. Comparative transcriptomics allows for the identification of genes that are developmentally, spatially or temporally controlled, or that are differentially expressed in response to different treatments or in distinct populations. This approach has allowed the functional characterization of many genes in model organisms, and next generation transcriptomics now permits adopting this approach in non-model organisms such as crops, wild relatives, and weedy species.

- For system-level analysis of complex traits, RNA-Seq provides information about complex interactions within biological systems. The response of the complete set of transcripts to a stimulus can be computationally analyzed to obtain an in-depth understanding of molecular biology systems. Owing to its holistic nature, genome-wide transcriptomics allows the identification of co-expressed genes, and the assembly of co-regulated gene clusters that are then used to build previously unpredictable gene regulatory networks. One of the ongoing goals for the system biology approach is to dissect the sophisticated genetic interactions that result in complex traits.

- Finally, transcriptome data can be used to accelerate marker-assisted breeding and genetic improvement in agriculturally valuable plants. Comparative transcriptomics aids in developing genome-wide *in silico* polymorphic genic markers such as SNPs and microsatellites. This application of RNA-Seq has great promise, for instance, in population genetic studies conducted on non-model plants.

RNA sequencing platforms: high quality-short reads vs. low quality-long reads

Researchers initiating transcriptome experiments will have to consider various popular technologies (e.g., Illumina MiSeq/GAIIx/HiSeq 2000 and PacBio) at NGS facilities. Key aspects to be considered are (i) the amount of sequences obtained per run, (ii) the cost per gigabase of sequence, (iii) the length and quality of reads and fragment size, and (iv) the error rate. The relative importance of each of these parameters will depend on the specific objectives of the project.

Overall, all platforms except PacBio yield short (approx. 150 bp) paired-end reads, for which the sequencing error rate is below 2%. The bottleneck for short read-based RNA-Seq technologies, conceptually as well as computationally, is that these reads have to be assembled into larger contigs, representing full-length transcripts. This is particularly relevant when working with non-model species where no sequenced genome is available as a reference. Assembly of short reads into contigs for transcriptomics is not as challenging as it can be for *de novo* genome assembly, because the length of the transcripts in most cases is limited to a few kilobases and because of the relatively low occurrence of repeated sequences in coding regions. In the past, the RNA-sequencing industry was dominated by the Illumina HiSeq 2000 platform. A run on this platform can yield a maximum of 600 Gb of approximately 150 bp paired-end reads with a very high accuracy (> Q30) and very low sequencing error rate (< 0.3%) (Quail et al. 2012).

In recent years, the revolutionary PacBio technology has emerged from Pacific Biosciences as an approach allowing the sequencing of full-length mRNA molecules. An evident advantage of this single-molecule sequencing technology for mRNA is that assembly is no longer required, because the full-length mRNA sequence is obtained without fragmentation. A major drawback, however, is its high sequencing error rate (> 12%) (Quail et al. 2012). In transcriptome studies, PacBio is becoming a key tool for splice variant analysis, as up to now assembling and quantifying the different isoforms produced from a single pre-mRNA based on short reads was too complex. The high sequencing error rate of PacBio sequencing is mitigated by the concomitant use of short reads from Illumina platforms. Similarly to *de novo* genome assembly, PacBio reads can provide the structure of the contig while Illumina reads ensure sequence accuracy. Although PacBio technology is currently mostly applied to model organisms and crops of agricultural interest, there is no doubt that the development of this technology, along with a decreased operating cost, will enable in-depth transcriptomics studies in non-model species.

RNA-Seq: Genome-wide identification of differentially expressed genes

Besides qualitative data on transcript sequences, RNA-Seq provides quantitative data on differential expression of transcripts with very high resolution and with a much lower detection limit than other techniques such as microarrays. Due to its digital nature, RNA-Seq has a linear detection dynamic range over five orders of magnitude. By contrast, accurate detection and quantification of low abundance transcripts using microarrays requires customization of the microarray platform in order to improve the density of the relevant probes. However, RNA-Seq does have an inherent bias towards longer transcripts (Oshlack and Wakefield 2009; Young et al. 2010).

Transcript abundances obtained from paired-end read sequencing are reported in Fragments Per Kilobase of transcript per Million fragments mapped (FPKM) (Trapnell et al. 2010). In the case of single reads, it is referred as Reads Per Kilobase of transcript per Million fragments mapped (RPKM) (Mortazavi et al. 2008). Transcript quantification is obtained from read counts but needs to be normalized to remove technical biases inherent in the sequencing technology, most notably the length of the transcripts and the sequencing depth of a run. These biases are corrected using algorithms that compute normalized FPKM/RPKM values. Comparative analysis of expression values for each transcript permits identification of differentially expressed genes between two or more samples.

One of the limitations of RNA-Seq for transcriptomic studies is the still relatively high cost, which often limits the investigations to a few samples. However, there is a need in modern biology for comparing the response of the transcriptome to many experimental conditions, or to compare the transcriptome of many plant species or populations. Multiplexing various mRNA samples into a single pooled sample is possible by ligating DNA tags with a specific index sequence for each individual sample. Therefore, sequencing several samples for the cost of a single run is possible, but the sequencing depth will be reduced as the number of pooled samples increases.

Real-time RT-PCR and microarray: Far from being obsolete

Validation of quantitative data is an important aspect of transcriptomics, and fully analyzed RNA-Seq data only constitute an entry point into the understanding of a molecular mechanism. Differential expression ideally has to be validated using different approaches and different samples to test both technical and biological reproducibility. Real-time RT-PCR is still the technique of choice to validate expression data for a relatively small number of genes. This can now be easily achieved in non-model organisms because

NGS platforms provide sequence information. For larger scale experiments, validation of gene expression can be achieved by either (i) running a second sequencing experiment or (ii) by using microarrays. In case of non-model organisms, specific microarray chips can be synthesized, taking advantage of the sequence data obtained by RNA-Seq.

As of today, microarrays are very reliable and more cost effective than RNA-Seq for gene expression in model organisms. Researchers will consider either genome-wide coverage in a few samples through RNA-Seq or a more limited gene analysis in a vast number of samples, depending on the research focus. In non-model organisms, the massive amount of data generated by genomic and transcriptomic projects should allow the development of species-specific microarrays. These microarrays will likely constitute major tools enabling cost-effective testing of many different samples and confirming expression data obtained from sequencing. The analysis of microarray results has become easier for inexperienced users, facilitated by the emergence of user-friendly software and free-of-charge analytical packages.

The complexity of RNA-Seq data analysis will soon be reduced as technology and bioinformatic tools progress. The cost of RNA-Seq will certainly drop over time as observed for microarrays. Microarrays can now be hybridized in personal stations and results can be analyzed without specific bioinformatics workflows giving most laboratories access to this technology. Although the sequencing industry is developing cheaper platforms enabling sequencing in individual laboratories (such as Illumina MiSeq), most of the sequencing activity is processed through core facilities. Another advantage of microarrays is that the amount of data generated is relatively small, which facilitates the handling, storage, and sharing of data files. There is no doubt that microarrays and RNA-Seq will remain complementary (see Fig. 3 for a comparison of published plant transcriptomics studies using microarrays and RNA-Seq). In the future, microarrays may be mostly relegated to massive comparative transcriptomic studies or used for neo-model organisms. RNA-Seq will eventually be used routinely as cost is decreased and data analysis is streamlined.

Applications of Transcriptomics for Weedy Species

Weed science has seen a burst of new experiments in the past few years utilizing the power of NGS to explore the transcriptomes of different weed species. The first two transcriptomes of weeds to be sequenced and assembled via RNA-seq methods were horseweed (Peng et al. 2010) and waterhemp (Riggins et al. 2010). Since then, the transcriptomes of 18 more weed species have been sequenced (Table 1), with the majority appearing in the last two years. Note that the transcriptomes reported here include

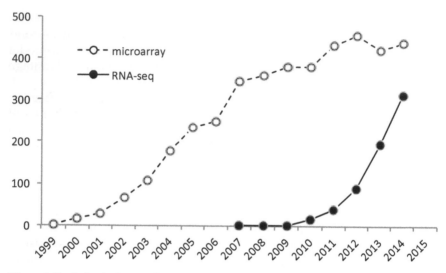

Figure 3. Evolution in the use of gene expression analyses. Number of published reports using either microarray or RNA-Seq for transcriptome analysis in plants (source: PubMed, using "microarray and plants" and "RNA-Seq and plants" as query. Data current as of September 23, 2015).

only those sequenced via NGS technologies. Prior to 2005, sequencing relied on Sanger technology which produced transcriptomes that contained only a portion of the total expressed genes and were generally not quantitative (Wang et al. 2009). For a more in-depth discussion about the study of weed transcriptomes before the advent of NGS, refer to the reviews by Lee and Tranel (2008) and Horvath (2010). With sequencing costs down to about $0.05/Mb (Wetterstrand 2015) and as weed scientists become increasingly familiar with this approach, it is likely that many more weed species will join this list in the coming years.

So far, the data mined from current transcriptomes are largely focused on herbicide resistance genes, including herbicide target-site genes (Riggins et al. 2010; Wiersma et al. 2015; Yang et al. 2013) and non-target-site resistance pathways (An et al. 2014; Doğramacı et al. 2015; Gaines et al. 2014; Gardin et al. 2015; Leslie and Baucom 2014; Peng et al. 2010; Riggins et al. 2010; Yang et al. 2013). However, many of the published transcriptome papers also explore the capacity of weeds to act as a repository of novel germplasm, providing breeders with new sources of desirable traits. Examples of these traits include insect, fungi, and herbivore resistance (Liu et al. 2014; Tsaballa et al. 2015), pod shatter resistance (Liu et al. 2014), C4 photosynthesis (Peng et al. 2014a; Yang et al. 2013), medicinal secondary compounds (Han et al. 2015), and potassium accumulation (Li et al. 2015). The potential for this work to contribute to the field of plant evolution is also well documented, encompassing studies of plant invasiveness (Huang et al. 2012), plant

Table 1. List of all currently available weed transcriptomes (as of September 2015).†

Scientific Name	Common Name	Data (Mb)	NCBI Accession(s)	Transcript #	Platform	Publication
Conyza canadensis L.	horseweed	109	PRJNA79565	31,783	454	Peng et al. 2010
Amaranthus tuberculatus	waterhemp	127	PRJNA79625	44,469	454	Riggins et al. 2010
Triphysaria versicolor	yellowbeak owl's clover	4775	SRX040930, SRX041138, SRX008134, SRX008135	67,794	454, IGA	Wickett et al. 2011
Striga hermonthica	witchweed	6009	SRX040928, SRX040929, SRX008132, SRX008133	61,539	454, IGA	Wickett et al. 2011
Phelipanche aegyptiaca	Egyptian broomrape	4893	SRX040924, SRX040925, SRX008130, SRX008131	51,479	454, IGA	Wickett et al. 2011
Mikania micrantha	bitter vine	1163	N/A	31,131	IGA	Huang et al. 2012
Echinochloa crus-galli	barnyardgrass	184	PRJNA170890	78,124	454	Yang et al. 2013
Thlaspi arvense	pennycress	37500	PRJNA183634	33,873	I2000	Dorn et al. 2013
Youngia japonica	Japanese hawkweed	4413	PRJNA221361	51,850	I2000	Peng et al. 2014a
Lolium rigidum	ryegrass	406	PRJNA239942	19,623	454	Gaines et al. 2014
Eleusine indica L.	goosegrass	23000	PRJNA239295	158,461	IGA	An et al. 2014
Sinapis arvensis	field mustard	8196	PRJNA232677	131,278	I2000	Liu et al. 2014
Ipomoea purpurea	morning glory	44000	PRJNA216984	65,459	IGA	Leslie and Baucom 2014
Cuscuta pentagona	dodder	23000	PRJNA217944	79,867	I2000	Ranjan et al. 2014
Kochia scoparia	kochia	103000	PRJNA239752	34,969	I2000	Wiersma et al. 2015
Alternanthera philoxeroides	alligator weed	43000	PRJNA268359	64,949	I2000	Li et al. 2015

Table 1 contd. ...

...Table 1 contd.

Scientific Name	Common Name	Data (Mb)	NCBI Accession(s)	Transcript #	Platform	Publication
Pueraria lobate	kudzu vine	9000	PRJDB2442	81,508	I1000	Han et al. 2015
Solanum elaeagnifolium	silver-leaf nightshade	4610	PRJNA284227	75,618	I2000	Tsaballa et al. 2015
Alopecurus myosuroides	blackgrass	127000	PRJNA234492	65,558	I2000	Gardin et al. 2015
Euphorbia esula	leafy spurge	54000	PRJNA243566	12,918	I2500	Doğramacı et al. 2015

†Abbreviations: Data, dataset size; Mb, megabases; Platform, NGS Sequencing Platform; 454, Roche/GS-FLX 454; IGA, Illumina GAIIx; I1000, Illumina HiSeq 1000; I2000, Illumina HiSeq2000; I2500, Illumina HiSeq 2500

parasitism (Ranjan et al. 2014; Wickett et al. 2011), and plant adaptation to wide environmental ranges (Yang et al. 2013).

Differential expression analysis in weeds

Transcriptomes are a rich resource for weed scientists because they provide the opportunity to study all of the expressed genes in an organism in one go. This is extremely useful for learning more about known genes of interest, including all versions of a particular gene, better gene models, posttranscriptional modifications, and quantification of absolute expression levels of a gene in a particular tissue/environment. Transcriptomic data are also useful for finding previously unknown genes that contribute to a phenotype, particularly for complex traits that involve more than one gene. By comparing differences in gene expression between two or more biotypes and/or across two or more environmental conditions, scientists can generate a list of candidate genes responsible for the phenotype in question. This type of experiment is commonly referred to as an RNA-Seq differential expression (DE) study, and Table 2 outlines the RNA-Seq DE experiments that have been conducted on weed species to date.

More than half of these studies have focused on using RNA-seq to understand the mechanism of resistance to one or more herbicides. One of the first differential expression experiments was carried out by Yang et al. (2013) on the grass species *Echinochloa crus-galli* (barnyardgrass), using NGS to identify both target-site and non-target-site genes involved in ALS inhibitor (penoxsulam and bispyribac-sodium) and synthetic auxin (quinclorac) resistance. ALS inhibitor resistance was also explored in two later studies, both of which were also conducted on grassy weed species. In

Table 2. Weed science differential gene expression experiments using a next generation sequencing platform.†

Scientific Name	Common Name	Application	Platform	DEA	Publication
Cirsium arvense	Canada thistle	Hybridization	IGA	In-house	Bell et al. 2013
Echinochloa crus-galli	barnyardgrass	ALS and synthetic auxin inhibitor resistance	454	DEGseq	Yang et al. 2013
Lolium rigidum	ryegrass	ACCase inhibitor resistance	I2000	DESeq	Gaines et al. 2014
Ipomoea purpurea	morning glory	EPSPS inhibitor resistance	IGA	EdgeR	Leslie and Baucom 2014
Cuscuta pentagona	dodder	Parasitism	I2000	EdgeR	Ranjan et al. 2014
Lolium rigidum	ryegrass	ALS inhibitor resistance	I2000	Modified DESeq	Duhoux et al. 2015
Alternanthera philoxeroides	alligator weed	Potassium stress	I2000	In-house	Li et al. 2015
Euphorbia esula	leafy spurge	Glyphosate tolerance	I2500	EBSeq	Doğramacı et al. 2015
Alopecurus myosuroides	blackgrass	ALS inhibitor resistance	I2000	DESeq	Gardin et al. 2015
Echinochloa crus-galli	barnyardgrass	Adaptive diversity	I2500	EdgeR	Nah et al. 2015

†Abbreviations: Application, Purpose of the RNA-seq study; Platform, NGS Sequencing Platform; 454, Roche/GS-FLX 454; IGA, Illumina GAIIx; I2000, Illumina HiSeq2000; I2500, Illumina HiSeq 2500; DEA, Differential Expression Analysis program used

the first study (Duhoux et al. 2015), a comparison of pyroxsulam resistant and sensitive ryegrass (*Lolium rigidum*) plants found 30 candidate non-target site resistant (NTSR) genes, four of which were validated in down-stream work. These four genes included two cytochrome P450s, one glycosyl-transferase, and one glutathione-*S*-transferase, confirming resistance was due to differential expression of one or more metabolic pathways. The second study (Gardin et al. 2015) looked at NTSR in iodosulfuron and mesosulfuron resistant blackgrass (*Alopecurus myosuroides* L.). Resistant and sensitive individuals from a segregating F_2 population were sequenced and compared to reveal five differentially expressed genes potentially involved in NTSR ALS inhibitor resistance, including three cytochrome P450s, one peroxidase, and one disease resistance protein.

Another RNA-Seq DE experiment was carried out in ryegrass, this time tracking down the genes responsible for ACCase inhibitor resistance (Gaines et al. 2014). Comparison of transcript expression levels between diclofop-resistant and -sensitive lines uncovered 28 genes that were constitutively over- or underexpressed in the resistant plants compared to the sensitive plants. Again, a majority of these genes were found to be metabolism-related and included three cytochrome P450s, a nitronate monooxygenase, three glutathione transferases, and a glucosyltransferase. Further validation experiments confirmed that expression changes in at least four of these DE genes were consistently associated with the resistance phenotype.

Non-target-site glyphosate resistance/tolerance has also been investigated via RNA-Seq approaches, both in morningglory (*Ipomoea purpurea*) and leafy spurge (*Euphorbia esula*). In the morningglory population, DE analysis found a medley of genes involved in metabolism, signaling, and defense to be differentially expressed between glyphosate-resistant and -sensitive biotypes (Leslie and Baucom 2014). In leafy spurge genes involved in shoot apical meristem maintenance, hormone biosynthesis, cellular transport, and detoxification were implicated (Doğramacı et al. 2015). These results are not surprising as glyphosate tolerance in leafy spurge is largely due to an induction of growth of underground adventitious buds after application of the herbicide.

Alongside these herbicide resistance-centered studies, RNA-Seq DE projects have been employed to study the advantageous genetic traits of certain weed species. A study by Ranjan et al. (2014) focused on the parasitic plant, dodder (*Cuscuta pentagona*), and found a number of genes that appeared to be responsible for the parasitic abilities of the plant, including genes involved in transport, response to stress, and cell wall modification. Another DE study conducted on alligator weed (*Alternanthera philoxeroides*) found a cohort of genes likely responsible for the plant's high tolerance for low potassium environments, revealing a network of genes that help the plant accumulate potassium and use it more efficiently (Li et al. 2015).

A few of the RNA-Seq projects conducted on weed species have focused on studying the genes that allow weedy species to invade and succeed in new environments. In 2015, the ability of barnyardgrass to grow readily across a range of climates was investigated by comparing the transcriptomes of multiple accessions grown under wet and dry conditions (Nah et al. 2015). The high adaptive diversity of these plants was found to be reliant on differential expression of upstream signaling pathways, including receptor-like kinase and calcium-dependent kinase genes. A separate experiment on Canada thistle (*Cirsium arvense*) found that the ability of intraspecific hybrids to survive and flourish in new environments was largely due to transgressive expression differences in the hybrids compared to either parent (Bell et al. 2013). Although the functional annotations of these transgressively over-expressed genes were not explored, this dataset is a useful resource for future work on heterosis.

The majority of the raw sequence data from these RNA-Seq DE studies have been published online and are available for future data mining efforts. As more accessions are added and their responses to varying environmental conditions are explored, we can expect to get a clearer picture of transcriptional regulation in weedy species.

Challenges of RNA-Seq for weed science

It is important to keep in mind that, while RNA-Seq is an excellent tool for weed scientists, there are some caveats to its use. First of all, RNA-Seq DE studies are hypothesis-generating studies. In order to reach robust conclusions about the system in question, follow-up work is needed to validate the results, including—at the very least—running real-time PCR to double-check the expression levels of select genes in the study population and unrelated populations. Secondly, phenotypic differences can arise due to reasons other than differential gene expression. Modifications to proteins can occur after protein translation, including protein phosphorylation or proteolytic cleavage that RNA-Seq is unable to identify. Other techniques like mass spectrometry or western blotting will be needed to detect these changes. Finally, the complex genomic landscape of plants makes studying their transcriptomes a daunting task at times. Plants are known to have much higher overall heterozygosity, ploidy level, and repetitive content compared to other eukaryotic genomes, leading to increased numbers of pseudogenes and larger gene families in general (Schatz et al. 2012). This, along with the high copy chloroplast and mitochondrial genomes, makes assembling a transcriptome difficult.

From Genomic Data to Novel Diagnostics and Applications to Herbicide Resistance Management

Herbicide resistance and diagnostics

Weed control in modern cropping systems is vital to protect crop yields, maintain profitable farming, and meet global food demands. Herbicides are major tools to control weeds and weed control failure caused by herbicide resistance is an increasing and significant problem worldwide (Heap 2015). Several factors, including the genetics and biology of the weed species, the herbicide chemistry and its Mode of Action, as well as key agro-ecosystem characteristics and herbicide handling by the users, influence the development of herbicide resistance which follows evolutionary processes (Christoffers 1999; Darmency 1994; Powles and Yu 2010). The evolution of herbicide resistance can occur rapidly when large and genetically variable weed populations are subjected to intensive herbicide selection (Powles and Yu 2010). Resistance to herbicides can result from several mechanisms occurring alone or in combination. The first group includes all modifications of the targeted proteins including gene coding sequence mutations, gene over-expression, and gene duplication, collectively known as Target Site Resistance (TSR). In the second group, processes not directly involving the herbicide target proteins such as the modification of the herbicide penetration into the plant, decreased rate of herbicide translocation, increased rate of herbicide sequestration, or enhanced metabolism are known as Non Target Site Resistance (NTSR) (Délye 2013). TSR confers to weeds a relatively narrow resistance to a single MOA or even subgroups of chemical classes within a MOA and is so far characterized by point mutations in the coding sequence (Heap 2015; Powles and Yu 2010) or gene duplication (Gaines et al. 2010) of the gene encoding the herbicide target. Related to enhanced expression of the herbicide target protein, no mutation has been described in promoter sequences or in related sequences, e.g., transcription factors. NTSR is a substantial threat, most often involving several genes, and especially in cases of herbicide detoxification (enhanced metabolic resistance or EMR), NTSR can confer resistance to a broad range of herbicides. The knowledge on NTSR mechanisms is still limited (Gaines et al. 2014). Quick, reliable and robust herbicide resistance diagnostics would offer the opportunity to select the most appropriate herbicide to control weeds with significant cost savings and benefit for the environment while avoiding application of the wrong treatment. Beffa et al. (2012) and Burgos et al. (2013) have recently reviewed methods for the detection and confirmation of herbicide resistance. Classical greenhouse bio-tests are still widely applied to detect herbicide resistance, but they require extensive greenhouse space and time (several weeks to 2 to 3 months) before an answer can be delivered. Molecular assays are now cheaper and more readily available.

Herbicide Target Site Mutations, Sequencing and Diagnostics

Recent development of sequencing technologies has allowed the application of pyrosequencing to detect TSR mutations and therefore diagnose herbicide resistance in days instead of weeks for the bio-tests. Nevertheless, only known mutations can be detected in a reasonable throughput and multiplex analyses, while possible, are not always reliable. Today, adaptation of Illumina sequencing technologies focused on a particular set can offer the possibility to detect several SNPs related to TSR in a high number of samples for reasonable costs. It would also be possible to detect new SNPs either in the coding region or in promoter regions of the gene encoding the herbicide target protein or in genes encoding any regulators of its expression (e.g., transcription factors). Each new SNP will have to be validated either by co-segregation with the herbicide-resistant phenotype or by functional analyses (e.g., gene expression, protein activity, or herbicide binding assay to the target protein).

Herbicide non target site resistance, genomics, data validation, and diagnostics

Non Target Site Resistance is thought to rely on the activity of multiple genes (Délye 2013; Powles and Yu 2010). Especially in species for which incomplete or no genomic sequences are available, comparisons of gene expression between herbicide-resistant and -sensitive populations have been performed using transcriptomic sequencing approaches (Table 2). Today NTSR resistance has been studied for a few herbicides and for a limited number of weed species (see section "Application of Transcriptomics in Weed Science"), mainly using 454 and/or Illumina sequencing. Usually several tens or hundreds of genes show differential expression, which for many is due to genetic variation between the sensitive and resistant populations and not due to NTSR. Therefore, one of the most critical questions is to identify what differentially expressed genes between resistant (R) and sensitive (S) plants are indeed correlated or involved in herbicide resistance and can be selected as validated molecular markers. First, to limit the number of differentially expressed genes, it is important to perform the transcriptome analyses on plants with genetic background as close as possible between R and S (e.g., Gaines et al. 2014). When this is not possible, a method to overcome this problem can be to analyze numerous populations represented either by several individuals or by mixtures of individuals. In this last case, in addition to gene expression, SNP analysis might be difficult. Once sequencing data are available, differential gene expression is generally validated by qPCR on the original cDNAs used for sequencing and/or cDNA from plants from other populations showing the same phenotype(s). Then genetic (e.g., co-segregation between gene expression level and resistance phenotype over at least 2 generations) and physiological validation (induction of R phenotype and correlation with gene expression level) have to be performed (e.g., Gaines et al. 2014). Finally these markers have to be tested in plants from field populations of different origins. This can be the minimum prerequisite validation to develop transcriptional diagnostic markers to detect NTSR. Ideally SNPs of a set of genes correlated with NTSR in combination with TSR SNPs would offer the opportunity to develop diagnostics based on high throughput sequencing and aimed to detect herbicide resistance related to multiple mechanisms. It should be noted that a strong correlation of transcriptional markers with NTSR does not necessarily mean that the products of the selected genes are directly involved in NTSR, in particular in the detoxification of the herbicide(s).

Reaching the conclusion of causation requires additional experiments aiming to characterize the function of the gene product either by over-expression or knock-out in transgenic model plants (e.g., Cummins et al. 2013), or analyzing the ability of the gene product to detoxify the herbicide in a heterologous system (e.g., yeast or bacteria), as most weed species cannot be transformed making functional genomics difficult to perform.

Application to Herbicide Resistance Management

Weed control relies on several technologies, including non-chemical (e.g., crop rotation, soil management) and chemical (herbicides). Weed resistance to herbicides is growing, and fast, reliable diagnostics offer the opportunity to choose the right product at the right time. Nevertheless multiple herbicide resistance mechanisms exist and this increases the complexity to develop diagnostics, likely requiring a set of multiple markers. Pyroseqencing as used today is a first step of such high throughput diagnostics allowing TSR based herbicide resistance to be assessed, but this is only part of the answer. Genomics and particularly sequencing technologies offer the unique opportunity to reveal and characterize key genes, and sometimes the associated SNPs, to be used as molecular markers for appropriate diagnostics enabling the best possible cost-effective and environmentally sound use of herbicides.

Conclusion

The continued increase in the scale and accuracy of NGS technology, along with the continually decreasing cost, is enabling genomic and transcriptomic research in weeds to proceed at an ever-increasing pace. Reference sequences can now be reasonably obtained for any species of interest, and bioinformatics tools needed to enable the use of NGS data are becoming increasingly available to biologists. Genomics approaches have enabled the study of various herbicide resistance traits, including traits with complex genetics such as enhanced herbicide metabolism. Future research will incorporate genomics approaches into the study of weed biology and enable the unraveling of complex genetic traits in weeds such as abiotic stress tolerance and competitiveness. In order to fully utilize the power of genomics in weedy species, additional experimental resources such as transformation systems, mutant lines, and genetic diversity panels will be needed. However, the increasing reality of the genomics era for weedy species offers exciting potential for improved understanding of what makes a weed a weed.

LITERATURE CITED

Agarwal P, Parida SK, Mahto A, Das S, Mathew IE, Malik N and Tyagi AK (2014) Expanding frontiers in plant transcriptomics in aid of functional genomics and molecular breeding. Biotechnol. J. 9: 1480–1492.

Albert VA, Barbazuk WB, dePamphilis CW, Der JP and Leebens-Mack J (2013) The Amborella Genome and the Evolution of Flowering Plants. Science 342: 1467–+.

An J, Shen X, Ma Q, Yang C, Liu S and Chen Y (2014) Transcriptome profiling to discover putative genes associated with paraquat resistance in goosegrass (*Eleusine indica* L.). Plos One 9: E99940.

Baird NA, Etter PD, Atwood TS, Currey MC, Shiver AL, Lewis ZA, Selker EU, Cresko WA and Johnson EA (2008) Rapid SNP discovery and genetic mapping using sequenced RAD markers. Plos One 3: e3376.

Banks JA, Nishiyama T, Hasebe M, Bowman JL and Gribskov M (2011) The selaginella genome identifies genetic changes associated with the evolution of vascular plants. Science 332: 960–963.

Beffa R, Figge A, Lorentz L, Hess M, Laber B and Ruiz-Santaella JP (2012) Weed resistance diagnostic technologies to detect herbicide resistance in cereal growing areas. A review. Julius-Kühn-Archiv 434: 75.

Bell GDM, Kane NC, Rieseberg LH and Adams KL (2013) RNA-Seq analysis of allele-specific expression, hybrid effects, and regulatory divergence in hybrids compared with their parents from natural populations. Genome Biology and Evolution 5: 1309–1323.

Birol I, Raymond A, Jackman SD, Pleasance S and Coope R (2013) Assembling the 20 Gb white spruce (Picea glauca) genome from whole-genome shotgun sequencing data. Bioinformatics 29: 1492–1497.

Blair AC and Hufbauer RA (2010) Hybridization and invasion: one of North America's most devastating invasive plants shows evidence for a history of interspecific hybridization. Evol. Appl. 3: 40–51.

Brenchley R, Spannagl M, Pfeifer M, Barker GLA and D'Amore R (2012) Analysis of the bread wheat genome using whole-genome shotgun sequencing. Nature 491: 705–710.

Burgos NR, Tranel PJ, Streibig JC, Davis VM, Shaner D, Norsworthy JK and Ritz C (2013) Review: confirmation of resistance to herbicides and evaluation of resistance levels. Weed Sci. 61: 4–20.

Burgos NR, Singh V, Tseng TM, Black H, Young ND, Huang Z, Hyma KE, Gealy DR and Caicedo AL (2014) The impact of herbicide-resistant rice technology on phenotypic diversity and population structure of United States weedy rice. Plant Physiol. 166: 1208–1220.

Burrell M, Pepper A, Hodnett G, Goolsby J, Overholt W, Racelis A, Diaz R and Klein P (2015) Exploring origins, invasion history and genetic diversity of *Imperata cylindrica* (L.) P. Beauv. (Cogongrass) using genotyping by sequencing. Plant and Animal Genome XXIII: W640.

Butler J, MacCallum I, Kleber M, Shlyakhter IA, Belmonte MK, Lander ES, Nusbaum C and Jaffe DB (2008) ALLPATHS: *de novo* assembly of whole-genome shotgun microreads. Genome Res. 18: 810–820.

Chen S, McElroy JS, Flessner ML and Dane F (2014) Utilizing next-generation sequencing to study homeologous polymorphisms and herbicide-resistance-endowing mutations in *Poa annua* acetolactate synthase genes. Pest Manag. Sci. 70. DOI:10.1002/ps.3897.

Christoffers MJ (1999) Genetic aspects of herbicide-resistant weed management. Weed Technol. 13: 647–652.

Cummins I, Wortley DJ, Sabbadin F, He Z and Coxon CR (2013) Key role for a glutathione transferase in multiple-herbicide resistance in grass weeds. Proc. Natl. Acad. Sci. USA 110: 5812–5817.

D'Hont A, Denoeud F, Aury J-M, Baurens F-C and Carreel F (2012) The banana (Musa acuminata) genome and the evolution of monocotyledonous plants. Nature 488: 213–+.

Darmency H (1994) Genetics of herbicide resistance in weeds and crops. *In*: Powles SB and Holtum JAM (eds.). Herbicide Resistance in Plants: Biology and Biochemistry. CRC Press, Inc., Boca Raton, FL.

Davey JW, Hohenlohe PA, Etter PD, Boone JQ, Catchen JM and Blaxter ML (2011) Genome-wide genetic marker discovery and genotyping using next-generation sequencing. Nat. Rev. Genet. 12: 499–510.

Délye C (2013) Unravelling the genetic bases of non-target-site-based resistance (NTSR) to herbicides: a major challenge for weed science in the forthcoming decade. Pest Manag. Sci. 69: 176–187.

Deschamps S, Llaca V and May GD (2012) Genotyping-by-sequencing in plants. Biology 1: 460–483.

Doğramacı M, Foley ME, Horvath DP, Hernandez AG, Khetani RS, Fields CJ, Keating KM, Mikel MA and Anderson JV (2015) Glyphosate's impact on vegetative growth in leafy spurge identifies molecular processes and hormone cross-talk associated with increased branching. BMC Gen. 16.

Dorn KM, Fankhauser JD, Wyse DL and Marks MD (2013) *De novo* assembly of the pennycress (*Thlaspi arvense*) transcriptome provides tools for the development of a winter cover crop and biodiesel feedstock. Plant J. 75: 1028–1038.

Duhoux A, Carrère S, Gouzy J, Bonin L and Délye C (2015) RNA-Seq analysis of rye-grass transcriptomic response to an herbicide inhibiting acetolactate-synthase identifies transcripts linked to non-target-site-based resistance. Plant Mol. Biol. 87: 473–487.

Eid J, Fehr A, Gray J, Luong K and Lyle J (2009) Real-time DNA sequencing from single polymerase molecules. Science 323: 133–138.

Elshire RJ, Glaubitz JC, Sun Q, Poland JA, Kawamoto K, Buckler ES and Mitchell SE (2011) A robust, simple genotyping-by-sequencing (GBS) approach for high diversity species. Plos One 6: e19379.

Gaines TA, Lorentz L, Figge A, Herrmann J and Maiwald F (2014) RNA-Seq transcriptome analysis to identify genes involved in metabolism-based diclofop resistance in *Lolium rigidum*. Plant J. 78: 865–876.

Gaines TA, Zhang W, Wang D, Bukun B and Chisholm ST (2010) Gene amplification confers glyphosate resistance in *Amaranthus palmeri*. Proc. Natl. Acad. Sci. USA 107: 1029–1034.

Garcia-Mas J, Benjak A, Sanseverino W, Bourgeois M and Mir G (2012) The genome of melon (*Cucumis melo* L.). Proc. Natl. Acad. Sci. USA 109: 11872–11877.

Gardin JAC, Gouzy J, Carrere S and Delye C (2015) ALOMYbase, a resource to investigate non-target-site-based resistance to herbicides inhibiting acetolactate-synthase (ALS) in the major grass weed *Alopecurus myosuroides* (black-grass). BMC Gen. 16.

Glaubitz JC, Casstevens TM, Lu F, Harriman J, Elshire RJ, Sun Q and Buckler ES (2014) TASSEL-GBS: A high capacity genotyping by sequencing analysis pipeline. Plos One 9: e90346.

Goff SA, Ricke D, Lan T-H, Presting G and Wang R (2002) A draft sequence of the rice genome (*Oryza sativa* L. ssp. *japonica*). Science 296: 92–100.

Goffeau A, Barrell B, Bussey H, Davis R and Dujon B (1996) Life with 6000 genes. Science 274: 546–567.

Grabherr MG, Haas BJ, Yassour M, Levin JZ and Thompson DA (2011) Trinity: reconstructing a full-length transcriptome without a genome from RNA-Seq data. Nature Biotech. 29: 644–652.

Han R, Takahashi H, Nakamura M, Yoshimoto N, Suzuki H, Shibata D, Yamazaki M and Saito K (2015) Transcriptomic landscape of *Pueraria lobata* demonstrates potential for phytochemical study. Frontiers in Plant Science 6.

Haudry A, Platts AE, Vello E, Hoen DR and Leclercq M (2013) An atlas of over 90,000 conserved noncoding sequences provides insight into crucifer regulatory regions. Nat. Genet. 45: 891–U228.

He J, Zhao X, Laroche A, Lu Z-X, Liu H and Li Z (2014) Genotyping-by-sequencing (GBS), an ultimate marker-assisted selection (MAS) tool to accelerate plant breeding. Frontiers in Plant Science 5: 484.

Heap I (2015) The international survey of herbicide resistant weeds. Available on-line: www. weedscience.com. Accessed March 25, 2015.

Holt RA, Subramanian GM, Halpern A, Sutton GG and Charlab R (2002) The genome sequence of the malaria mosquito *Anopheles gambiae*. Science 298: 129–149.

Horvath D (2010) Genomics for weed science. Current Genomics 11: 47–51.

Huang YL, Fang XT, Lu L, Yan YB, Chen SF, Hu L, Zhu CC, Ge XJ and Shi SH (2012) Transcriptome analysis of an invasive weed *Mikania micrantha*. Biologia. Plantarum 56: 111–116.

Hutchison CA (2007) DNA sequencing: bench to bedside and beyond. Nuc. Acids Res. 35: 6227–6237.

JGI Genome Portal (2016) The Genome Portal of the Department of Energy Joint Genome Institute. Available at: http://genome.jgi.doe.gov/viridiplantae/viridiplantae.info.html.

Korlach J, Marks PJ, Cicero RL, Gray JJ and Murphy DL (2008) Selective aluminum passivation for targeted immobilization of single DNA polymerase molecules in zero-mode waveguide nanostructures. Proc. Natl. Acad. Sci. USA 105: 1176–1181.

Lander ES, Linton LM, Birren B, Nusbaum C and Zody MC (2001) Initial sequencing and analysis of the human genome. Nature 409: 860–921.

Lee RM and Tranel PJ (2008) Utilization of DNA microarrays in weed science research. Weed Sci. 56: 283–289.

Lee RM, Thimmapuram J, Thinglum KA, Gong G, Hernandez AG, Wright CL, Kim RW, Mikel MA and Tranel PJ (2009) Sampling the waterhemp (*Amaranthus tuberculatus*) genome using pyrosequencing technology. Weed Sci. 57: 463–469.

Leslie T and Baucom RS (2014) *De novo* assembly and annotation of the transcriptome of the agricultural weed *Ipomoea purpurea* uncovers gene expression changes associated with herbicide resistance. G3-Genes Genomes Genetics 4: 2035–2047.

Levene MJ, Korlach J, Turner SW, Foquet M, Craighead HG and Webb WW (2003) Zero-mode waveguides for single-molecule analysis at high concentrations. Science 299: 682–686.

Li L, Xu L, Wang X, Pan G and Lu L (2015) *De novo* characterization of the alligator weed (*Alternanthera philoxeroides*) transcriptome illuminates gene expression under potassium deprivation. Journal of Genetics 94: 95–104.

Liu J, Mei D, Li Y, Huang S and Hu Q (2014) Deep RNA-Seq to unlock the gene bank of floral development in *Sinapis arvensis*. Plos One 9: E105775.

Lu F, Lipka AE, Glaubitz J, Elshire R, Cherney JH, Casler MD, Buckler ES and Costich DE (2013) Switchgrass genomic diversity, ploidy, and evolution: novel insights from a network-based SNP discovery protocol. PLoS Genet. 9: e1003215.

Luo R, Liu B, Xie Y, Li Z and Huang W (2012) SOAPdenovo2: an empirically improved memory-efficient short-read *de novo* assembler. Gigascience 1: 18.

Margulies M, Egholm M, Altman WE, Attiya S and Bader JS (2005) Genome sequencing in microfabricated high-density picolitre reactors. Nature 437: 376–380.

Maughan PJ, Smith SM, Fairbanks DJ and Jellen EN (2011) Development, characterization, and linkage mapping of single nucleotide polymorphisms in the grain smaranths (*Amaranthus* sp.). Plant Genome 4: 92–101.

Mayer P, Farinelli L and Kawashima EH (2011) Method of nucleic acid amplification. U.S. Patent No. 7,985,565.

Michael TP and Jackson S (2013) The first 50 plant genomes. The Plant Genome 6. DOI:10.3835/plantgenome2013.03.0001in.

Mortazavi A, Williams BA, McCue K, Schaeffer L and Wold B (2008) Mapping and quantifying mammalian transcriptomes by RNA-Seq. Nature Methods 5: 621–628.

Myers EW, Sutton GG, Delcher AL, Dew IM and Fasulo DP (2000) A whole-genome assembly of *Drosophila*. Science 287: 2196–2204.

Nah G, Im J-H, Kim J-W, Park H-R, Yook M-J, Yang T-J, Fischer AJ and Kim D-S (2015) Uncovering the differential molecular basis of adaptive diversity in three *Echinochloa* leaf transcriptomes. Plos One 10: e0134419.

NCBI Genome (2016) National Center for Biotechnology Information, U.S. National Library of Medicine. Available at: http://www.ncbi.nlm.nih.gov/genome.

Neale DB, Wegrzyn JL, Stevens KA, Zimin AV and Puiu D (2014) Decoding the massive genome of loblolly pine using haploid DNA and novel assembly strategies. Genome Biol. 15.

Nystedt B, Street NR, Wetterbom A, Zuccolo A and Lin Y-C (2013) The Norway spruce genome sequence and conifer genome evolution. Nature 497: 579–584.

Okada M, Hanson BD, Hembree KJ, Peng Y, Shrestha A, Stewart CN Jr, Wright SD and Jasieniuk M (2013) Evolution and spread of glyphosate resistance in *Conyza canadensis* in California. Evol. Appl. 6: 761–777.

Oshlack A and Wakefield MJ (2009) Transcript length bias in RNA-seq data confounds systems biology. Biol. Direct 4: 14.

Peng Y, Abercrombie LLG, Yuan JS, Riggins CW, Sammons RD, Tranel PJ and Stewart CN (2010) Characterization of the horseweed (*Conyza canadensis*) transcriptome using GS-FLX 454 pyrosequencing and its application for expression analysis of candidate non-target herbicide resistance genes. Pest Manag. Sci. 66: 1053–1062.

Peng Y, Gao X, Li R and Cao G (2014a) Transcriptome sequencing and *de novo* analysis of *Youngia japonica* using the Illumina platform. Plos One 9: E90636.

Peng Y, Lai Z, Lane T, Nageswara-Rao M and Okada M (2014b) *De novo* genome assembly of the economically important weed horseweed using integrated data from multiple sequencing platforms. Plant Physiol. 166: 1241–1254.

Pfeifer M, Martis M, Asp T, Mayer KFX, Lübberstedt T, Byrne S, Frei U and Studer B (2013) The perennial ryegrass Genome Zipper: Targeted use of genome resources for comparative grass genomics. Plant Physiol. 161: 571–582.

Poland JA, Brown PJ, Sorrells ME and Jannink J-L (2012) Development of high-density genetic maps for barley and wheat using a novel two-enzyme genotyping-by-sequencing approach. Plos One 7: e32253.

Powles SB and Yu Q (2010) Evolution in action: Plants resistant to herbicides. Annu. Rev. Plant Biol. 61: 317–347.

PubMed (2015) National Center for Biotechnology Information, U.S. National Library of Medicine. Available at: https://www.ncbi.nlm.nih.gov/pubmed.

Quail MA, Smith M, Coupland P, Otto TD, Harris SR, Connor TR, Bertoni A, Swerdlow HP and Gu Y (2012) A tale of three next generation sequencing platforms: comparison of Ion Torrent, Pacific Biosciences and Illumina MiSeq sequencers. BMC Gen. 13: 341.

Ranjan A, Ichihashi Y, Farhi M, Zumstein K, Townsley B, David-Schwartz R and Sinha NR (2014) De novo assembly and characterization of the transcriptome of the parasitic weed dodder identifies genes associated with plant parasitism. Plant Physiol. 166: 1186–+.

Rensing SA, Lang D, Zimmer AD, Terry A and Salamov A (2008) The Physcomitrella genome reveals evolutionary insights into the conquest of land by plants. Science 319: 64–69.

Riggins CW, Peng YH, Stewart CN and Tranel PJ (2010) Characterization of *de novo* transcriptome for waterhemp (*Amaranthus tuberculatus*) using GS-FLX 454 pyrosequencing and its application for studies of herbicide target-site genes. Pest Manag. Sci. 66: 1042–1052.

Ronaghi M, Uhlén M and Nyrén P (1998) A sequencing method based on real-time pyrophosphate. Science 281: 363–365.

Ronaghi M, Karamohamed S, Pettersson B, Uhlén M and Nyrén P (1996) Real-time DNA sequencing using detection of pyrophosphate release. Analytical Biochemistry 242: 84–89.

Rothberg JM and Leamon JH (2008) The development and impact of 454 sequencing. Nature Biotech. 26: 1117–1124.

Sanger F, Nicklen S and Coulson AR (1977) DNA sequencing with chain-terminating inhibitors. Proc. Natl. Acad. Sci. USA 74: 5463–5467.

Schatz MC, Witkowski J and McCombie WR (2012) Current challenges in *de novo* plant genome sequencing and assembly. Genome Biol. 13.

Schneeberger K, Ossowski S, Lanz C, Juul T, Petersen AH, Nielsen KL, Jorgensen J-E, Weigel D and Andersen SU (2009) SHOREmap: simultaneous mapping and mutation identification by deep sequencing. Nature Methods 6: 550–551.

Schulz MH, Zerbino DR, Vingron M and Birney E (2012) Oases: Robust *de novo* RNA-seq assembly across the dynamic range of expression levels. Bioinformatics 28: 1086–1092.

Sonah H, Bastien M, Iquira E, Tardivel A and Legare G (2013) An improved genotyping by sequencing (GBS) approach offering increased versatility and efficiency of SNP discovery and genotyping. Plos One 8: e54603.

The 1001 Genomes Consortium (2016) 1135 genomes reveal the global pattern of polymorphism in *Arabidopsis thaliana*. Cell 166(2): 481–491.

The Arabidopsis Genome Initiative (2000) Analysis of the genome sequence of the flowering plant *Arabidopsis thaliana*. Nature 408: 796.

Trapnell C, Williams BA, Pertea G, Mortazavi A, Kwan G, van Baren MJ, Salzberg SL, Wold BJ and Pachter L (2010) Transcript assembly and quantification by RNA-Seq reveals unannotated transcripts and isoform switching during cell differentiation. Nature Biotech. 28: 511–515.

Tsaballa A, Nikolaidis A, Trikka F, Ignea C, Kampranis SC, Makris AM and Argiriou A (2015) Use of the *de novo* transcriptome analysis of silver-leaf nightshade (*Solanum elaeagnifolium*) to identify gene expression changes associated with wounding and terpene biosynthesis. BMC Gen. 16.

Tuskan GA, Difazio S, Jansson S, Bohlmann J and Grigoriev I (2006) The genome of black cottonwood, *Populus trichocarpa* (Torr. & Gray). Science 313: 1596–1604.

Wang Z, Gerstein M and Snyder M (2009) RNA-Seq: a revolutionary tool for transcriptomics. Nat. Rev. Genet. 10: 57–63.

Watson JD and Crick FHC (1953) Genetical implications of the structure of deoxyribonucleic acid. Nature 171: 964–967.

Wetterstrand K (2015) DNA sequencing costs: Data from the NHGRI Genome Sequencing Program (GSP). http://www.genome.gov/sequencingcosts. Accessed: September 21, 2015.

Wickett NJ, Honaas LA, Wafula EK, Das M and Huang K (2011) Transcriptomes of the parasitic plant family Orobanchaceae reveal surprising conservation of chlorophyll synthesis. Curr. Biol. 21: 2098–2104.

Wiersma AT, Gaines TA, Preston C, Hamilton JP, Giacomini D, Buell CR, Leach JE and Westra P (2015) Gene amplification of 5-enol-pyruvylshikimate-3-phosphate synthase in glyphosate-resistant *Kochia scoparia*. Planta 241: 463–474.

Wu GA, Prochnik S, Jenkins J, Salse J and Hellsten U (2014) Sequencing of diverse mandarin, pummelo and orange genomes reveals complex history of admixture during citrus domestication. Nature Biotech. 32: 656–+.

Xu X, Pan S, Cheng S, Zhang B and Mu D (2011) Genome sequence and analysis of the tuber crop potato. Nature 475: 189–U94.

Yang X, Yu X-Y and Li Y-F (2013) *De novo* assembly and characterization of the barnyardgrass (*Echinochloa crus-galli*) transcriptome using next-generation pyrosequencing. Plos One 8: E69168.

Young MD, Wakefield MJ, Smyth GK and Oshlack A (2010) Method Gene ontology analysis for RNA-seq: accounting for selection bias. Genome Biol. 11: R14.

Yu J, Hu S, Wang J, Wong GK-S and Li S (2002) A draft sequence of the rice genome (*Oryza sativa* L. ssp. *indica*). Science 296: 79–92.

Zimin A, Stevens KA, Crepeau M, Holtz-Morris A and Koriabine M (2014) Sequencing and assembly of the 22-Gb loblolly pine genome. Genetics 196: 875–890.

Index

Milton Keynes UK
Ingram Content Group UK Ltd.
UKHW051015071024
449327UK00012B/259